Wild
MOMS

ALSO BY DR. CARIN BONDAR

Wild Sex

Wild

MOMS

MOTHERHOOD *in the* **ANIMAL KINGDOM**

DR. CARIN BONDAR

PEGASUS BOOKS

NEW YORK LONDON

Wild Moms

Pegasus Books Ltd.
148 W. 37th Street, 13th Floor
New York, NY 10018

First Pegasus Books paperback edition October 2019

First Pegasus Books hardcover edition April 2018

Interior design by Maria Fernandez

Library of Congress Cataloging-in-Publication Data is available.

ISBN: 978-1-64313-232-7

Printed in the United States of America
Distributed by W. W. Norton & Company

*I dedicate this book to my amazing, supportive,
and wonderful mom, Joanne*

*. . . and to those silly people who call me mom:
Shaeden, Loanna, Faro, and Juna*

CONTENTS

INTRODUCTION

I t could be argued that motherhood is the most important job that evolution ever made. And yet what it means to be a mom is something that is both impossible to tangibly state and totally obvious to anyone who has done it. We can learn some lessons about our own species by looking at our animal counterparts, and that's why, as an evolutionary biologist and a mother of four myself (with children who are 12, 10, 8, and 6), I've spent a great deal of time studying motherhood in the animal kingdom. The result is *Wild Moms*. It's been an extraordinary journey, and the thing that continually surprised me throughout my research has been that, for all the differences between distantly related animals, what is most fascinating is how much we all have in common.

One of the most compelling things I considered as I researched this book is how human mothers vary in our practices as natural "animals." After all, our bodies evolved to survive and reproduce, and the former certainly doesn't mean much without the latter. In my own way, I somehow just knew when I was ready for motherhood. It happened fairly fast—and interestingly, prior to that point, I really wasn't sure when or if I would even have children. It's almost like a switch went off inside my body and brain, and suddenly I could not get pregnant fast enough. I was obsessed with babies, getting pregnant, and learning what I could do to facilitate that process. The strange way in which my body "decided" I was ready for motherhood always left me wondering whether any other animal moms out there feel the same way. Although as observational scientists we aren't yet in a position to simply ask animals about their feelings and emotions about various aspects of their lives, recent research on the emotional and cognitive capabilities of animals allows us to conclude that there is most definitely a level of awareness and sensitivity in many members of the animal kingdom. Animals form relationships, they empathize, they love. What could possibly be more significant for any female (mammal, fish, or bird) than to undertake the process of motherhood?

Now that my kids are a little older, I'm able to reflect more on my experience and what it means in the context in the rest of the animal kingdom. Let's face it: for those first few years, any human mother is running off her feet just trying to ensure that her infants survive. There are countless feedings, changes, fussiness, and tremendous exhaustion. It's so difficult, and yet supposedly so natural. Well, it turns out that while much of our experience matches that of our animal counterparts, human mothers are guilty of complicating the process to extents that are unprecedented in the animal kingdom. We will learn in this book that while modern technologies and Western lifestyles allow for an increased level of comfort and support, there is simply no level of wealth or materialism that can replace what the body is naturally meant to do. Women can be amazing mothers without all the bells and whistles that many modern practices require. Just because chimpanzee mothers do it without bottles, strollers, formula, or diapers doesn't mean that they are any less maternal than human mothers. In fact, there are many cases in which I would consider our lavish developments to make our mothering practices less natural than those of the rest of the animal kingdom—although that's a topic for another book.

Wild Moms sees the reader through the process of motherhood, from the unfertilized egg to becoming a grandmother. In all chapters, I highlight examples of the ways that animal moms address the same kinds of issues that human moms have. This book is of course far from exhaustive; a complete volume of motherhood could (and should) be many times longer than this. However, you can rest assured that I've included a wealth of information that is interesting, educational, and poignant.

Overall, I hope that this book opens your eyes to the diverse and amazing world of motherhood on our planet. I've spent so much time on this topic, yet sometimes I feel as though I have barely scratched the surface. My hope is that *Wild Moms* will be an exciting introduction to the simultaneous diversity and sameness of motherhood. My overall conclusion is one that I've always carried with me when it comes to raising my own children: That there isn't just one way to be successful. If we define success as the survival of an infant to sexual maturity, then there are as many ways to be successful as there are moms out there doing it. In most animals, including primates, this is where our "measurement" of motherhood largely stops. In our own species, however, it seems that we have taken this issue and made it into a new kind of problem—that of "infants" who remain dependent long after sexual maturity takes place. I suppose that's a topic for another book, but suffice it to say, that human moms have a lot more in common with the rest of the primate order than we sometimes like to think.

Thank you for your interest, and happy reading!

Wild
MOMS

soup of sperm and eggs all madly trying to reach the correct partners at the correct time.

Another factor that broadcasters must take into account is density of potential partners and potential competitors. Conditions can change at a moment's notice, and triumphant individuals will be those who are able to process information in a timely manner. Too many or not enough spawners could spell disaster for an individual's reproductive success. Eggs that do not come into contact with the appropriate sperm will go unfertilized; however, eggs that come into contact with too much sperm could be susceptible to polyspermy (multiple fertilization), which renders them useless. Although broadcasting sounds like an effective way to get the job done, it could be detrimental for individual reproductive success if gamete densities become too high.

Many broadcast spawning mothers are also fathers. Hermaphrodites are common in the invertebrate world, and a large proportion of aquatic invertebrates have both male and female reproductive organs. That's not to say that these individuals can simply fertilize their own gametes and not participate in sex with others. There are species where this is possible, but it's generally better for genetic diversity to cross-mate with either one or more outside individuals, in the same way that humans avoid incest to avoid the potential negative effects of recessive alleles (genetic abnormalities that are normally masked through outbreeding). In a pinch, one may fertilize oneself if the environment isn't conducive to successful fertilization with an outside partner, but it's always better in terms of biological fitness to diversify the genetic lineage so as to hedge your bets that at least some of the offspring will survive.

Genetic diversity is indeed an important aspect of reproduction for all mothers, but in several species females can produce genetic clones of themselves asexually, in a process called parthenogenesis or thelytoky. While this process does not require a genetic contribution from a male (sperm), most parthenogenetic females can also reproduce sexually when the opportunity presents itself. In fact, a female may choose a complex course of reproductive options depending on the environmental cues that she experiences at a given time. When would a female choose to simply clone herself over accepting a genetic donation from a male partner? In some cases, males may not be available, or it may be wise for a female to reproduce in the fastest way possible due to external factors like predation risk. Being able to clone one's self is a pretty interesting way to be able to maximize one's biological fitness in the face of environmental stochasticity (unpredictability), sexually transmitted diseases, or immediate predation risk.

In addition to having the ability to clone themselves through asexual reproduction, the female springbok mantis (a terrestrial invertebrate closely related to other more recognizable praying mantis species) also engages in cannibalism of their potential male partners. In this species females cannibalize males *before* any sexual activity. In other words, males are often nothing more than a light snack for hungry females who reserve the ability to clone themselves instead of mating with them. It's a bit of a double whammy for males of this species—not only is their sperm on a take-it-or-leave-it basis with potential female partners, but they also stand to lose their lives even before they get a chance to share it.

It turns out to be the age of females that is the most likely predictor of their behavior toward males. These ladies become less aggressive and more likely to mate sexually as they age. This is a common theme in many sexually cannibalistic females. Initially, they get their bodies in prime mothering condition by ingesting a few sexually mature males. Reproducing asexually through pathogenesis allows these females to cannibalize males at a high frequency while they still maintain the ability to pass on their genetic blueprints. However, later in life when their bodies are in prime condition for reproducing sexually, females switch reproductive modes. This scenario is described by the *adaptive foraging hypothesis*, which predicts a decrease in cannibalism as females age and an increase in fecundity of the females who engage in it.

Females who can reproduce asexually aren't always the ones in the driver's seat. There are certain bacterial or viral endosymbionts (organisms that live and reproduce exclusively within another) that can force clonal reproduction in their hosts. This is common in arthropods (which are segmented invertebrates including insects and arachnids). The endosymbionts work to increase their own biological fitness by inducing asexual reproduction in their female hosts. Can you imagine if your own body just reproduced without any kind of decision to do so on your part? This is the unfortunate reality for over 124 species in seven various arthropod categories. The bacterial genus *Wolbachia* is the culprit, causing unsuspecting females across the world to become asexual mothers without their knowledge or consent. Females in question aren't harmed by this lack of control over their (asexual) reproductive lives; they maintain the ability to reproduce sexually when an appropriate opportunity presents itself.

Broadcast spawning and parthenogenesis are two ways that female animals become moms, but they aren't the routes that are normally associated with the process. As mentioned above, viviparity loosely translates to "alive reproduction," while oviparity refers to egg reproduction. Whether offspring are produced through viviparity or oviparity, a male cannot be sure of his paternity; but in species in which eggs are laid early in development and require tending, protection, or incubation, he may make a substantial paternal investment. This is because dads can do something to help since the embryos are *outside* of the mother's body. (That of course is not the case in most mammalian species where the egg is internally fertilized.)

There is an extensive array of ways in which oviparity takes place in the animal kingdom. In some species, mothers construct elaborate nests, burrows, or hiding spots for egg incubation; in other species, embryos are attached to the mother's body in a specialized pouch or in a crevasse on her body, or they are otherwise carefully guarded until they hatch. The type and extent of egg care is largely dictated by the ecology of the species in question and by the order in which egg extrusion and fertilization occur. If fertilization comes before egg-laying, females lay embryos rather than eggs. Many invertebrate, fish, and amphibian mothers lay egg cases filled with fertilized embryos after they copulate with males and receive sperm. However, if fertilization occurs after the fact, females lay eggs (not embryos) that then receive sperm from nearby males. The number of eggs or embryos depends on the mother's strategy for pre-hatching care. If she knows she won't be around, she may lay hundreds or thousands of eggs to increase the chance that some of them will develop into adulthood. If she or the father is going to be able to stick around and care for the potential offspring, there will usually be fewer.

One important aspect of egg-case biology is that not all offspring are created equally. Mothers provide some embryos within the egg case with nutritional advantages; neglected embryos become nothing more than a snack for their own brothers or sisters. It's *Game of Thrones: Egg Case Edition*, in which some embryos are destined for survival and others destined to die horrific deaths at the hands of their siblings. For example, a few species of marine snails engage in something called *accidental cannibalism*. Occasionally, early developing embryos ingest younger embryos, which continue to develop once they are in the first embryo's gut. The ingested embryo eventually bursts free from its cannibalistic sibling, killing it in the process. It can then engage in something called *oophagy*, or the ingestion of its sibling's leftover parts.

Terrestrial invertebrates also lay egg clutches. These aren't necessarily encapsulated together but instead are laid close together in the same area. For example, female flies of various species lay egg clutches in microhabitats according to environmental cues like resource availability or competition from other females. Eggs can be concentrated onto specific resource patches—like a pile of dung, a tree hole, a specific plant, or a dead carcass—that seem like they might present some developmental advantages. These moms fine-tune both the number and sex of the embryos that they deposit to maximize the eggs' own individual probabilities for reproductive success.

The most straightforward strategy is simply overproducing offspring without providing specific advantages to any of them. Mothers then let their offspring duke it out for survival within their own egg-case groups. In this way, the strongest, fittest offspring survive and have an excellent chance of making it to sexual maturity themselves. Indeed, females often lay more than ten times the number of eggs that survive to adulthood. These embryos—who had been forced to fight for their lives before they had a chance to be born—are a far cry from the completely helpless newborns of our own species.

It may seem somewhat heartless for a mom of any kind to create more than enough offspring, but there are several well-founded hypotheses for why this may occur. The first reason is explained by the *siblicide advantage hypothesis*, in which the strongest and largest (often the first hatched) embryos cannibalize their own siblings and realize the twofold benefit of less overall competition and increased body mass and nutrition. The *predator satiation hypothesis* suggests that moms who lay more than enough offspring stand a better chance at having at least a few of them survive a predator attack. Lastly, the *insurance policy hypothesis* suggests that mothers may create more offspring so that they can be biologically covered in the event of an environmental disturbance or hatching failure. It's important to keep in mind that most species exhibit some kind of flexibility in terms of the specific number of embryos or strategy that can be used in a given situation. Unlike most mammals, many invertebrates display a remarkable level of reproductive variability.

As I mention above, in many species, females and males engage in copulation *prior* to the laying of already-fertilized eggs, whereas in other species, females lay eggs that are externally fertilized by males after the fact. This is a key point in the ecology of many fish species exhibiting the latter, as females may simply take off and leave their eggs to a male to fertilize and then care for. For example, in some bony fish species like salmon and trout, mothers will dig out small nests or redds in benthic (bottom) areas of various marine and freshwater environments. They use their tails to excavate depressions in the substrate, into which they then deposit their eggs. Some females will remain with their eggs once they are fertilized; however, most will leave all egg guarding and maintenance to the father. Since males are often defending a territory, females can choose where to lay their eggs based on qualities of the territory, or on characteristics of the male himself. In many species females are more likely to deposit their eggs into the nest of a male that already has some—since this indicates that other females have chosen him, he's likely a good provider.

In fish species that are externally fertilizing, males ordinarily take over the bulk of embryonic care once females have left their eggs behind. However, there are always exceptions to any generalization in the animal kingdom. For example, female mouth-brooding cichlids in Lake Tanganyika do not leave their eggs to the care of a specific male. Instead, once they have extruded their eggs, they scoop them back up into their own mouths. Females then solicit sperm donations from various potential fathers. Since it's best for a female to maximize the genetic diversity of her offspring, she may seek the sperm of several males for fertilization. Females keep the fertilized embryos in their mouths for the duration of their gestation, and new hatchlings will continue to seek protection in mom's mouth for a short time after they hatch. Although it's a strange sight to imagine (a female accepting sperm into her mouth from several males in sequence), a mom's gotta do what a mom's gotta do!

External fertilization and egg-guarding by males is more often the rule in fish. Although egg laying takes place in some of these species too, in the cartilaginous fishes (such as sharks and rays) copulation, internal gestation, and viviparity are more common, and it's in these species that we expect to see the greatest level of maternal care. *Chondrichthyans* (cartilaginous fishes) have a diverse set of reproductive modes, with strategies that range from egg laying to bearing of live young to physiologically complex placental development and nutrition of developing embryos. Shark reproduction is notoriously difficult to study, and sexual reproduction varies substantially between genera. For example, female sand tiger sharks exhibit a remarkable reproductive mode that combines elements of both oviparity and viviparity; this hybrid form of reproduction is called "ovoviviparity."

Females give birth to live young that have been nourished by egg yolks inside the uterine cavity. There is no nutritive connection between mother and babies once the fertilized eggs are laid within her own body, which is quite different from the direct nutritive connections (e.g. placenta) that occur in purely viviparous species. Female sand tiger sharks have a pair of uteri, and embryos travel to one of the females' uteri where they remain for a further period of gestation after being fertilized. However, the first embryo in each uterus to reach 2.1 to 2.3 inches in length hatches and then begins a period of exponential growth as it systematically attacks, kills, and cannibalizes all its siblings in that uterine environment. Once the siblings have all been ingested and the hatchling has reached a size of approximately 4 inches in length, it will begin to feast on the unfertilized ova that have accumulated in the mother's uterus. Imagine all this action taking place within one's body while one simply carries on with the day-to-day tasks of life. A veritable war is taking place within her uteri, and female sharks need only wait for the fittest and strongest offspring to emerge at birth. They will be approximately 37½ to 50 inches at this point and will have few predators for the rest of their lives. What this means for sand tiger shark moms is that the strongest, fittest juveniles will triumph from a war that takes place before her babies are even independent. Moms can seek sperm donations from a variety of potential fathers, but only the sperm of the very fittest males will be the successful sire of her offspring. In this way, a female can be behaviorally polygynous but functionally monogamous from each uterus.

Egg-laying is thought to be the ancestral form of creating offspring through sexual reproduction, and most vertebrates continue to utilize it in a number of diverse ways. However, the bearing of live offspring has evolved more than 150 times in the vertebrate groups (apart from birds and jawless fishes), making it one of the greatest examples of convergent evolution. Viviparity has evolved independently in mammals, reptiles, amphibians, and fish. By contrast, lactation evolved once; endothermy and powered flight evolved three times. What does this mean exactly? It means that there are many benefits of internal gestation and live birthing. Such biological benefits have allowed for its repeated, independent evolution in several animal groups. Live birth is advantageous because it can provide enhanced offspring survival, the ability to exploit novel ecological niches, and an increase in the energetic efficiency of the entire reproductive process.

Viviparous vertebrates provide energy to developing offspring in a variety of ways. Matrotrophy, for example, is the provision of nutrition from a mother via a placenta or other means such as through secretory tissues or glands that allow for fetal absorption or ingestion of nutrients. In addition to nutrition derived from direct matrotrophy, there is also something called *lecithotrophy*, in which nutrition comes from within the egg. This can happen in both egg-laying and live-bearing animals; egg-laying animals are exclusively lecithotrophic, whereas viviparous animals can be both lecitotrophic and matrotrophic if embryos receive at least some of their nutrition from an egg yolk during their internal gestation.

Matrotrophy can be further divided into six categories. There is *oophagy* (in which sibling yolks or products of siblings are used as nutrition), *embryophagy* or *adelphophagy* (in which cannibalism of siblings occurs), *histophagy* (in which nutrients are ingested in liquid form), *macrophagy* (in which the mother's own tissues are used as offspring nutrition), *histotrophy* (in which nutrients travel from mom to offspring via diffusion), and *placentotrophy*, (in which nutrients are supplied to offspring through a placenta). A given animal mom is likely to make use of more than one of these kinds of processes, either at the same time or in succession. Since there have been so many independent evolutionary origins of these processes, it goes without saying that there is more than one way to successfully gestate an offspring. All mammals are highly matrotrophic, and that will be addressed in a greater level of detail in the *buns in mammalian ovens* chapter. But there are a few other animal groups that illustrate the immense diversity of what it means to become a mom.

Amphibians are another group of moms with a hugely variable reproductive repertoire. Not only can many amphibian moms manipulate the specific number of embryos involved in a reproductive event, but they can manage the specific ways in which they develop. Amphibians are usually characterized by an aquatic larval stage (think tadpoles), but, in the case of the fire salamander, some females forego the tadpole stage and give birth to directly developed terrestrial offspring, making them pueriparous. The larvae of many invertebrate species can be either direct developers that hatch as mini-adults or indirect developers with an aquatic larval stage. Fire salamanders show a remarkable level of plasticity when it comes to the specifics of gestating their offspring. In fact, populations frequently exhibit two co-occurring strategies for reproduction based on their specific ecological situation (i.e. it's possible for one individual to deploy either strategy depending on what will bring the most success). These strategies are called *pueriparity* (the birth of fully developed, terrestrial juvenile salamanders) and *larviparity* (the live-birth of aquatic larvae). Essentially, pueriparous mothers forego the larval stage and directly give birth to mini-adults, much like in mammals.

Some amphibian species may perhaps realize a greater level of reproductive success through bypassing the aquatic larval stage altogether. Larviparous females ovulate between twenty and eighty eggs that are fertilized through sexual copulation; they then start to develop synchronously. During this phase of development, the embryos feed on yolk stores from their eggs. Females give birth directly into the water, and the larvae subsequently develop for several more months before emerging onto land as terrestrial adults. The situation is different in pueriparous females, in which not all ovulated eggs are fertilized. Unfertilized eggs are arrested and provide food to early developing embryos. This strategy is common to all pueriparous salamanders. On the surface, it looks like a brood reduction; however, once one begins to investigate, it becomes much more than a simple reduction in numbers. In addition to the fact that not all eggs are fertilized, not all fertilized embryos develop at the same rate within the female's uterus.

The "lucky" embryos that develop the quickest can consume whatever else they can find within their mother's uterus. After approximately three months, the mother gives birth to between one and fifteen terrestrial juveniles, a much smaller number than the number of tadpoles produced by larviparous moms. However, the remaining embryos that manage to survive to parturition have a greater chance of survival than any individual tadpole. The evolution of live larval birth in amphibians is rarer than it is in other vertebrate groups, but it marks a very interesting prospect for their evolution. With time, pueriparity could eventually evolve into a new strategy where these salamanders can leave behind their aquatic habitat and colonize new habitats. Clearly, pueriparity is an example of the evolutionary process in progress, however, it's still quite rare in salamanders a whole, occurring in only 11 of 661 species.

There are some parallels between pueriparous salamanders and the egg cases produced by the marine snail and polychaete mothers. In both cases, a mother's investment into a cohort of offspring is unequal between individuals. Instead of all embryos developing at an equal pace with roughly the same amount of nutrition, there is a vastly unbalanced allotment of nutrition that produces larger or faster-developing individuals and smaller or slow-developing nurse eggs that are destined to help the former. In the salamanders, this could represent an evolutionary transition to direct development (i.e. foregoing the aquatic tadpole stage altogether), and a similar trend is apparent in the invertebrates (marine snails and polycheate worms). Mothers who produce egg cases with relatively equal investment into each embryo produce planktotrophic larvae that are free swimming. Planktotrophic larvae then settle to the substrate (the ocean bottom, a rock, a shell, or any relatively static structure) to metamorphose into adults. Mothers who produce egg cases with *unequal* provisioning to the embryos inside (i.e. nurse eggs for extra nutrition to fast developing individuals) may have offspring that bypass the planktotrophic stage and emerge from their egg cases as mini-adults. Perhaps an extra boost of nutrition in early development can facilitate an evolutionary transition to direct development. It's incredible to think that greater nutrition could essentially lead to skipping an evolutionary step; the implications of that are astounding.

Amphibians also exhibit something called *pseudoviviparity*, which is an interesting offshoot of oviparity, or egg-laying. Embryos are secondarily internalized by females after they are externally fertilized by males. The female reproductive system is therefore not directly involved in the gestation or the fertil-ization of the embryos in pseudoviviparous species. It sounds somewhat ridiculous to describe it in this way, so perhaps a few examples will do a better job of communicating the meaning of pseudoviviparity.

A great example comes from the gastric brooding frogs, which live in the tropical aquatic environments of Australia. Females swallow early-stage larvae and brood them in their stomachs for the duration of their development. Essentially, the stomach becomes a makeshift uterus (the actual reproductive system is not involved), and the offspring eventually make up almost 70% of mom's entire mass. When it's time to "give birth," some females lie passively and allow the infants to crawl up their digestive tracts and out through their mouths; the more active expel the fully developed froglets through a process called propulsive vomiting. (Yes, they actually vomit out their own babies; talk about a tremendous feat of motherhood.)

The male sexual partners of pseudoviviparous Surinam and Pipa pipa toads place fertilized eggs on the mother's dorsal surface where they are embedded in a matrix of skin and epithelium that completely covers them up. When the toadlets are ready to hatch, they break out of their mother's back, which again is not technically the same as a viviparous birth but most definitely merits substantial maternal investment.

As the previous section attests, the physiological processes that result in motherhood are vastly diverse. It's always humbling to think about the diversity of the animal kingdom, and the perspective we can gain from accepting that the "same" job can be accomplished in a multitude of ways—even within a species. Having an open-minded approach to diversity in both forms and functions of motherhood will serve you well as you progress through the chapters of this book.

2

THE EGG-CONTENT CONUNDRUM

t's difficult for human moms to think about "playing favorites" among our children. We aim instead to provide relatively equal opportunities, resources, and love to all our children. During the gestational phase, we attempt to keep our bodies well-fed and rested to provide a healthy environment for our developing infants. Moms of other species do so as well, but they don't have prenatal yoga and vitamins and pregnancy pillows to help them get a good night's sleep. In the unforgiving natural world, conditions can change quite suddenly, rendering critical resources unavailable. Another important factor to consider is the identity of a given father. Females can be forced to copulate, or may "settle" for a lesser-equipped suitor if practical choices are limited. Can any of these things lead to unequal investment in siblings of the same mom? Can mom somehow skew the gestational habitat so that it is more favorable to specific offspring if they have a high-quality dad or the environmental conditions are highly conducive to their survival? Can her body somehow preferentially provide increased nutrition or immunity to some babies over others?

The complexities of internal gestation frustrate measurement of this kind of phenomenon in mammals, but mothers of the bird world make it easy for us to observe. Eggs are little packages of maternal investment that provide a bounty of information about the specific ecological expectations of each developing chick. In addition to DNA from both mom and dad, eggs contain extra stuff provided exclusively by their mother—proteins, fats, vitamins, hormones, antioxidants, and antimicrobials to name just a few—which in turn provides biologists with a detailed glimpse into the intricate complexity of something termed *maternal effects*.

Many factors provided to infants during early gestation (including hormones and immune compounds) play major roles not just in early embryonic development, but also far into juvenile and even adult stages. Therefore, the specifics of how a mother builds her eggs yields important insight into the factors that maximize biological fitness of both mom and baby. Sometimes, the optimal strategy for both mother and child are the same; often they are not.

Many aspects of egg composition can be directly manipulated by mothers prior to laying, from major traits such as size, yolk volume, shell color, and overall egg weight to the minute concentrations of steroids and hormones, nutrients, sugars, vitamins, immune compounds, and antioxidants. There is an immense complexity to the egg laying business across the bird world, and the major questions surrounding the allocation of maternal effects have to do with why mothers would vary specific compounds in specific contexts.

The answers are immensely heterogeneous, and not necessarily exclusive of one another.

Egg size correlates to embryonic survival; individuals from larger eggs have a higher chance of surviving than those from smaller ones. This makes intuitive sense, as large eggs obviously contain larger amounts of "stuff" for juveniles to make use of. The advantages of coming from a large egg are realized not just early in life but also into adulthood, as individuals that hatch from larger eggs possess advantages for growth that continue throughout their lives. It's illuminating to look at the diversity in egg size within the egg clutches, since there can be major differences between sibling eggs from a single hatching bout. Most bird moms lay two or more eggs at a time, and size is often a function of the order in which they were laid, with the largest coming first and progressively getting smaller. The fact that later eggs are smaller can be due to several contributing factors; one of the more interesting scenarios involves the action of something called *brood reduction*. Brood reduction is exactly as it sounds—moms actually lay *extra* eggs that aren't necessarily meant to make it too far past the hatching or juvenile stage of life. These are analogous to the nurse eggs discussed in the last chapter on modes of reproduction. There are both facultative and obligate brood reducers (i.e. those that sometimes reduce and those that always do it). Those in the former category can utilize brood reduction as a strategy when the immediate ecological circumstances warrant it; however, they can also keep their entire brood if resources are abundant and even later-laid eggs can be amply provisioned. Environmental, social, and ecological factors can all potentially play a role in determining the fate of late-laid eggs. Unfortunately for last-laid eggs of obligate brood reducers, the only reason they were laid in the first place is as insurance against the unlikely fail of those that were laid before. If the early-laid eggs of obligate brood reducers hatch and rear successfully, these insurance chicks are always killed.

Facultative brood reducing moms may or may not oversee the death of their later-laid chicks, which means that the scene is set for a complex interplay of behaviors and personalities within the nest. It's *Lord of the Flies: Bird Nest Edition.* When resources are plentiful, or at least adequate, mothers can improve their own biological fitness the most if all their chicks survive to fledging. This may conflict with the biological desires of first-born dominant chicks, since they would prefer to share with as few siblings as possible to maximize their own biological fitness. Dominant chicks can be very aggressive toward junior chicks, pecking them, biting them, preventing them from getting food brought back to the nest, ejecting them from the nest, or killing them. In many bird species, moms give junior chicks a better chance at being able to fight back against older aggressive siblings by providing more nutrition via bigger yolks or by providing them an increased concentration of male-based hormones called androgens within the egg.

The role of androgens in the eggs is vast and variable according to context. Studies have shown conflicting results regarding the specific benefits and drawbacks of androgen provision; it's likely to be linked to other characteristics and effects may be species specific. An extra shot of androgens is usually helpful for junior chicks to develop more aggressive behaviors. This increased aggression is helpful for protecting themselves against their older siblings, and it can also provide them with a bolder, louder approach to begging for food. Parents often feed the loudest and most boisterous chicks in the nest, and the extra hormones provided to the junior chicks go a long way to helping them find their begging voices. This kind of "hormonal favoritism" is generally observed in species in which chicks aren't siblicidal or highly violent toward one another. Canaries and blue-footed boobies are two examples of species where competition between siblings can be intense but non-siblicidal. Moms will feed their junior chicks more efficiently when resources allow for it, and when they don't, as during resource poor conditions, the junior chicks will likely meet their untimely fate because they are prohibitively small and without an extra surge of hormonal favoritism.

But why would a mom ever lay an extra egg if there's a chance it won't survive and she will be left to facilitate its death? It's a fair question, and one that comes up often in the unforgiving natural world. In the case of egg laying and the potential for unpredictable environments, moms can use a bet-hedging strategy to maximize their genetic blueprints in future generations. In many species, the physiological mechanisms of egg laying dictate that a certain number of eggs will be laid after copulation. For example, in blue-footed boobies, females always lay two or three eggs—regardless of the environmental conditions. During each bout of laying, a female will make the choice as to how to provision each egg for the best possible outcome. If resources are poor, she will likely lay the subsequent eggs without extra provisioning, creating a small egg with low hormones to minimize the costs to her own biological fitness. Unprovisioned junior chicks will die quickly, and this will allow the mom to proceed with caring solely for her surviving, dominant chick. However, when resource conditions are good, it makes biological sense for a mom to make a larger investment in her second and third eggs, such that the junior chicks have a better shot at competing with the older siblings and surviving to the fledging stage. Mothers might also lay more eggs than will survive, but they aim to keep the number of surviving chicks as high as possible according to the available resources. For the bluefooted-boobies, they can have one surviving offspring when times are bad, and the chance for two surviving offspring when times are good. Since egg investment is a small allotment of a female's lifetime energy budget, it usually makes sense for her to at least try for another offspring.

Similarly, yellow-legged gull moms will lay two eggs, and, if food is plentiful, they will lay a third egg. The third egg is usually smaller than the first two, an adaptation of brood reduction in which the siblings will likely kill the third chick. Moms can also increase the size of their third egg, but they only do so if the chick inside is female. Daughters usually hatch at a size disadvantage to sons regardless of hatching order, so a chick that is both female and last-hatched will be at a double disadvantage. Mothers make up for this by giving the last-born female eggs a bit of extra albumen, which is high in protein and directly linked to skeletal growth. In this way, mothers can attempt to level the playing field for all chicks when the resource levels permit them to do so.

Female gulls are more susceptible to later hatching while male chicks are more susceptible to sibling competition. Male chicks are more aggressive toward each other than they are toward their sisters, and so male-on-male aggression can be a major contributing factor to brood reduction. The number of surviving chicks in an entire brood is higher if there are more females within the brood—and this is related not only to the less aggressive nature of female gull chicks, but also to their overall smaller size: They need less food, so there is less direct competitiveness to obtain it. Mothers can vary the sex ratio of chicks in their broods to maximize their own biological fitness depending on a multitude of characteristics at any given time. Resource levels play a major role, but it's just the first step when it comes to sex ratio, albumen provisioning, and hatching order.

For the most part, there aren't scenarios in which good conditions become suddenly bad or the reverse. Instead, there is likely going to be a gray area of resource availability somewhere in the middle. What then? Interestingly, there are times when it can work out best for blue-foot moms if they have two chicks, and each plays a distinctive role. When resources are moderate and survival of two chicks is possible, the best brood for a blue-foot mom may be one moderately aggressive chick and one subordinate chick. Both chicks can be sustained until fledging with minimal competition between them since their behavioral roles are clearly defined. This makes nest provisioning easier and more predictable. Mom can facilitate development of these two behavioral phenotypes by laying the second egg four days after the first and with approximately 10% less yolk volume. In this way, mom has set the scene for her junior chick to survive, but in a clearly defined submissive role.

The scenario is quite different for a closely related species: the brown booby. These are obligate brood reducers, meaning that moms always allocate fewer resources to their last-laid eggs since they will not be directly contributing to her biological fitness. Brown booby moms lay their last egg at least five days later than all other eggs to ensure that the junior chick is smaller at hatching. Siblicide is the unavoidable result. If the first chick hatches and proves its ability to thrive, it's lights out for the junior chick. Cattle egret moms do something similar by providing extra resources and steroids to their two first-laid eggs and almost none to the third. A whopping 85% of third-laid chicks of this species are killed by their stronger siblings. Maximizing one's biological fitness can be a cruel process indeed.

In the siblicidal black-legged kittiwakes the core young are socially dominant and large, and the marginal young are subdominant and smaller. In kittiwakes, the core young will fatally attack the marginal young when resources are poor. There is virtually no chance of a marginal chick becoming a dominant chick in this species, regardless of maternal provisioning. Their fate is determined solely by their laying order. In this species, the eggs of marginal young are also supplemented with extra androgens, making the marginal chicks quite aggressive, but they will not become dominant chicks within their own nests.

Biologists have hypothesized that this higher level of aggression may work to their evolutionary advantage. Kittiwakes are colonial breeding seabirds that establish large groups on ocean-side cliffs. Nests often sit in rows along the stratified horizontal rock layers, which can mean that nests each take up a position on a massive vertical grid on a cliff face. Adoption occurs at a high rate in this species, as nestlings have a fairly good chance at landing in a different nest if they happen to fall out (or get kicked out) of their own. Since marginal chicks are either killed or ejected from the nest by their dominant siblings, they stand a better chance of surviving in a foreign nest if they are more aggressive against their new siblings and exhibit frequent and enthusiastic begging behavior—both of which are characteristic of chicks hatched from androgen-supplemented eggs. The best outcome for a marginal chick that gets ejected from its own nest by a dominant sibling would be to land in a nest with hatchlings that are smaller or of similar size. The marginal chick could potentially become the new dominant chick if it manages to eject the smaller chicks and assume a dominant position in its new nest. This of course assumes that the parents of the foreign nest either do not notice or do not mind the extra foreign chick that has suddenly appeared. But in these kinds of species, adoption is quite common. Called the *adoption facilitation hypothesis*, this idea suggests that a kittiwake mom can increase her biological fitness with her marginal, later-hatching chicks by provisioning them with androgens. The androgens provide the marginal chicks with aggressive behaviors that make no difference in their natal nests but can serve them well in adoptive nests. It's an ingenious kind of evolutionary insurance.

The discussion of obligate brood reduction is interesting for a number of reasons, not the least of which is why such a scenario would evolve in the first place. There are many ways to answer this question for the various species that exhibit obligate brood reduction; however, brood reduction may also mitigate the damages caused by "brood parasitism," in which the mothers of one bird species deposit their eggs in the nests of a different species (discussed in greater detail in the next chapter). The depositing mom (the parasite) will often purposefully damage the eggs of her host species at the same time, a double whammy for the host mom who is not only providing care to eggs and offspring that are not even her own species, but who now has had her own eggs damaged or tossed from the nest.

Obligate brood reduction is observed in many species that are subject to high levels of brood parasitism. There are a few good reasons why these phenomena are related. First, if a mom produces extra eggs within her nest, she is more likely to recognize foreign (i.e. parasite) eggs. If there are several eggs of one size or color and just one that is different, this can help the mom to identify it as a parasite egg and remove it from her nest. In addition, if a host mom lays extra "insurance" eggs of her own, this could compensate for the damage or eviction of some of them. This is called the *replacement offspring hypothesis*, and it is supported by data from many tropical bird species affected by brood parasitism, such as the Montezuma, which is parasitized by the giant cowbird. Montezuma females maximize their reproductive success by employing a combination of resistance behaviors, like rejection of the parasite eggs, and compensation behaviors, like laying an extra egg to compensate for the potential loss to parasites. In species that demonstrate support for the replacement offspring hypothesis, the value of surplus eggs in terms of their weight, size, and provisioning is high. This means that any egg can hatch into a high-quality chick, and those that do not hatch are likely to have been damaged or evicted by parasitic moms. The laying of extra eggs—and the development of an obligate brood reduction strategy—is likely to be an important defense against the devastating effects of brood parasitism for many species.

Mother birds are also able to alter the unique properties of their eggs based on the identity of the father. Paternal quality can play a huge role in a mother's investment in his eggs. After all, not all male partners were created equally, and there is a myriad of ways in which females discern their most appropriate partner. The mating systems employed in the bird world are diverse—and include varying levels of paternal care and investment. There are species in which fathers provide little more than a sperm donation and others in which fathers provide a high level of parental care. When females have mated with a high-quality male, they are more likely to provision their eggs to reflect it. This is beautifully demonstrated by a study of blue-footed booby moms whose egg-laying was (rudely) interrupted by biologists interested in the nature of maternal investment and male quality.

In this species, the sexually selected characteristic is the bright, blue-green feet of the male. This is a fluid characteristic that is a reliable indicator of a male's health and viability, as well as a signal of his overall level of potential for paternal care. Since females generally lay two or three eggs (each approximately seventy-two hours after the other), biologists were able to modify a female's egg provisioning by changing the status of her mating partner during the interval between the first and second eggs. They did this by painting his bright blue feet a dull gray-brown. This obvious and important change to the "fitness" of her male partner caused the female to lay second eggs that were smaller in size and had a lower antioxidant content. The changes observed during this experiment were significantly different from what would have been normally expected for a second-laid egg, indicating that females are tuned in to the quality level of their mating partners.

Physical characteristics of males are often a primary factor in their level of attractiveness in birds; peacocks are a perfect example. However, they are not the only means by which females can tell a good mate from a bad one. For Eurasian reed warblers, it's all about the song. Females prefer more complex songs from their male partners, and they alter their breeding behavior and egg investment according to song attractiveness. Aspects of songs including rate, versatility, song type, or repertoire size can all influence the size of individual eggs as well as the overall number of eggs laid. Song complexity is positively associated with territory tenure, immune system quality, longevity, and lifetime reproductive success.

The level of testosterone provided to the eggs also varies according to the song characteristics of their fathers. Testosterone has direct physiological effects on muscle development and behavioral effects including an increase in frequency and intensity of begging behavior. And by giving her male eggs increased testosterone when her mate has an attractive song, a female can influence her son's chances of becoming an attractive mate. This is essentially a long-term increase to her own biological fitness, since she is increasing her chances of becoming a grandmother. Lastly, female warblers provision eggs of attractive dads with increased levels of lysozyme, an antimicrobial compound that is costly to produce. Lysozyme provides a protection to developing embryos within the egg and for a short period after hatching. This provides the chick with an increased chance of surviving to adulthood.

Both zebra finch and canary moms increase the androgen content of their eggs when their mate is highly attractive versus when he is not, and this boost of androgen has effects that circle back to dad himself. Increased levels of androgens in eggs have varying results on offspring fitness and behavior depending on the specific bird species being studied, but, in both these species, the extra androgens play a role in developing more aggressive behavior. The most attractive males in a population are often associated with a lower level of parental care, so the extra bump in androgens could be a way for moms to boost the level of begging behavior in the chicks that may serve to attract their father's attention. In this way, an attractive dad equals more androgens, which leads to increased begging effort, which results in the father paying greater attention. Females can also potentially manipulate egg components, egg size, and egg coloration—all to attract more interest from dad.

The *sexually selected egg-color hypothesis* suggests that the hue of blue-green eggs of many avian species may be the result of such a scheme. The color is due to a pigmentation from an antioxidant called biliverdin, which signals a superior female condition. It's a way of females letting high-quality males know that they are up to the male's standards and thus the eggs deserve extra attention. This is seen in starlings, robins, and house wrens, which all use aspects of egg coloration to have a positive impact on the level of male care. Females are thus able to use egg characteristics to manipulate paternal involvement. One can't help but wonder about the possibilities if women were able to take steps to ensure their male partners would be on the ball for midnight feedings and dirty diaper changes.

In addition to being a form of manipulation of males, females can change egg contents in response to environmental factors. For example, when population density is high and a female's chicks are likely to experience a high level of competition, moms often provision their eggs with extra androgens or testosterone to ensure that their offspring will be able to hold their own. When predator levels are predictably high, moms can provision their eggs with components for increased growth rates or faster wing development—all in the name of ultimately maximizing her own genetic representation in future generations.

Social system is another major factor in determining egg allocation levels, especially when the social system is one of cooperative breeding, or group-living species, in which one or a few dominant mating pairs produce most of the offspring. Non-mating adults remain with the group as "helpers" and do not generally attempt to breed. Helping adults are often close relatives of the mating pair, such that the help that they provide contributes to their own indirect biological fitness through kin selection. There are complicated rules for breeding and social dominance in cooperatively breeding societies, but the bottom line is that dominant breeding females in cooperative groups reliably have extra help to raise their offspring. If a helper female produces offspring of her own, she will not have the extra helping resources that are generally available to the dominant female.

Social status of moms in cooperatively breeding societies makes a big difference in terms of how eggs are provisioned. The *load-lightening hypothesis* describes a situation in which a dominant mom with helpers intentionally lays smaller, less provisioned eggs. Since she has the luxury of outside help to take care of her chicks, she doesn't need to expend the full energy necessary to give her children every advantage. Instead, she puts herself in better shape for survival by investing less in her offspring and allowing the helpers to make up for it. This is observed in several species of birds called sociable weavers.

Smaller eggs, however, do not result in smaller chicks at fledging. The extra food provided by adult helpers means that moms can get away with a measurably lower investment in their eggs. For these weavers, it's not simply a matter of egg mass either. The smaller eggs also contain lower amounts of testosterone, androgens, and corticosterone, the latter of which is a hormone that is passively transferred to eggs based on mom's stress level. In other words: moms with lots of help are less stressed, and their eggs reflect this. This should not be a shock to any new mom, human or animal.

It's of note that sociable weavers live long lives, up to sixteen years. This is meaningful in the context of a mother's breeding investment over just one season. Any time a mom can get by with relying on helpers to support her eggs, it would make biological sense for her to decrease her own investment in current offspring and save it for either her own survival or a future reproductive bout. Why give 100% when it's not necessary? The sociable weaver moms utilizing the help of others provide a clear example of a tradeoff between current and future reproduction, and it's fascinating that such tremendous "insight" can be translated to minute changes to eggs within a clutch.

Cooperatively breeding moms also consider the current and near-future availability of food. The primary question for any cooperatively breeding mom to consider is this: if environmental conditions are challenging, is it worth the risk of providing very little to eggs? How much can or should a mom rely on helpers to come through for the sake of a brood that isn't their own?

When environmental conditions are predictably favorable, such as when the surrounding area is cool and lush with plenty of available food sources, dominant breeding fairy-wren moms lay much smaller eggs and realize a benefit to their overall biological fitness for a comparatively small investment. On the other hand, when the environmental conditions are unpredictable or unfavorable, such as when climate is hot and dry with a concurrent decrease in available food sources, dominant breeding moms maintain a high level of investment, producing clutches with large eggs. This indicates that fairy-wren females are aware of the increased risk of relying on helpers when conditions are rough, and they act accordingly. This is evidence for the *bigger-is-better hypothesis*, which predicts that mothers can benefit by producing offspring (eggs) that are larger when conditions are potentially more threatening. Larger chicks are generally less vulnerable to external conditions like heat stress or evaporative weight loss, and so the chances of their survival in unfavorably hot and dry conditions is much higher than small chicks.

Compared to the dominant mothers with helpers, mothers without helpers do not appear able to make such fine-tuned changes to their eggs. Since these moms are doing all the foraging and chick provisioning on their own, they do not have the energy to be able to produce larger eggs when environmental conditions become less favorable. And even during favorable environmental conditions, creating smaller, less-fit eggs isn't a great idea, as these moms don't have any extra help to care for their future offspring. Overall, dominant moms in cooperatively breeding scenarios seem to have the upper hand when it comes to biological fitness. The conundrum of this situation for individuals that do not happen to find themselves in a dominant position is very real.

Bird moms of several species show a fine-tuned ability to micromanage the contents of their eggs, which is a fascinating demonstration of the evolutionary process at work. The forces of natural selection (i.e. survival of the fittest) can act at any stage of development, and bird moms play a major role in manipulating the biological destiny of each chick they produce. Far from just being something we eat for breakfast, eggs provide a unique look at the spectacular control that mothers have over the development of their offspring.

3

BROOD PARASITISM:
AN INCONVENIENT TRUTH

Imagine undertaking all the work of constructing and laying eggs only to abandon them—for good—in a foreign nest. Incredulous as it sounds, this kind of behavior is common in a large number of bird species that are called *brood parasites*. Moms make sure that their eggs are kept safe and warm through to hatching, not by doing this themselves but instead by depositing their eggs into the nests of other females of a different species.

The act of throwing away one's babies to the care of another mother who is not only unrelated but is also of a completely different taxonomic group must seem counterintuitive to any mother. However, many avian moms realize a great deal of biological fitness through this unusual strategy. Brood parasites are birds that do not brood their own eggs or tend to their own offspring once those eggs hatch. Brooding and provisioning offspring are instead left to mothers (and fathers) of another species.

There are actually *two* mothers to think about in the brood parasite scenario: the parasite mom and the host mom. Unfortunately, there really isn't much of a "bright side" for host moms, who do all the work for zero biological benefit. Not only do they put in a great deal of effort in taking care of infants that are of a different species, but parasite chicks often employ strategies that render them more successful than the host mom's own chicks. It's essentially a case of bullying: parasite chicks may push host chicks out of the nest, or they may beg louder and more frequently and hence obtain a greater level of parental provisioning. Some parasite moms will also poke holes in the eggs of host moms, so that the host mom's own chicks die an early death while the parasite eggs hatch and gain all her attention.

Brood parasites are of two kinds: generalists, which leave their eggs in the nests of a wide variety of other birds, and specialists, which deposit their eggs into the nests of only a very specific partner species. Generalist parasites like common cuckoos deposit their eggs in nests of a range of other birds. Although they may have a higher rate of rejection, they can be reproductively successful in a range of environments and under a variety of conditions. It's a bet-hedging strategy that works out successfully over the reproductive lifetime of any given female. Biologists predict that generalist brood parasitic species may thrive despite human-caused changes in climate or landscape because of their flexibility; they can lay eggs virtually anywhere, at any time. At the opposite end of the spectrum are the specialist brood parasites, which deposit their eggs into the nests of a very specific host species. In specialist parasite scenarios, there is a closer arms race between host and parasite moms. For example, specialist eggs may be virtually identical to those of host eggs, requiring host moms to evolve intricate mechanisms to be able to identify their own offspring or their own eggs. The chances of specialized-parasite breeding success in any one nest is very high, much higher than for a generalist parasite; unlike the generalists, however, they must live where their host species lives.

Because the maternal strategies of parasite and host moms are directly opposing, the brood parasite system (especially that of specialist brood parasites) is a model system for coevolution. Generalist and specialist parasite moms aim to do the best by their own offspring by finding the perfect host. Think of a busy mother's endless interviews for a top-tier nanny for her child. The maternal investment for brood parasitic moms consists of a suite of mechanisms that serve to make sure that her babies end up in a good home. For example, great spotted cuckoo moms parasitize the nests of magpie moms in many temperate areas of Europe. The volume of the magpie nests is related to the potential quality of the nest for a brood parasite in a few ways. First, nest volume is known to be an indicator of territory quality (larger nests are built in higher quality territories). Second, the timing of breeding in magpies is also related to territory quality, with those in high-quality territories breeding earlier than those in low quality ones. Naturally, parasitic moms preferentially aim to deposit their eggs into the high-volume nests of host moms in high-quality territories. After all, if you're going to give up your offspring to the care of a mother of a different species, there should be some measures in place to ensure that the host mother will be of high quality.

The nest is of utmost importance. Parasite moms really scope out their options when identifying suitable nests, taking into consideration the age of the host mom, the vocal repertoire of the host mom, and the microhabitat of the potential host nest. A certain level of "imprinting" occurs in parasitic chicks when they are newly hatched that causes them to seek out nests with characteristics similar to those that they experienced when they were young, including nest site, habitat, and parental characteristics. Many generalist brood parasites preferentially select the nests of naïve host moms who are more likely to accept their eggs as their own. However, parasitic moms cannot solely depend on the likelihood of finding naïve moms, so generalists have a repertoire of strategies to employ in various situations.

Once the nest has been selected, it's a question of getting the egg in into it. Cowbird and great spotted cuckoo generalist parasites deposit their eggs in nests that have a high-level of host-parental activity. This strategy is aptly termed the *host-activity hypothesis*, and it is a means by which parasite moms can avoid depositing eggs in nests that have been abandoned or that have a low level of parental provisioning. The great spotted cuckoo makes every effort to lay their eggs into appropriate nests, especially high-volume, high-parental activity nests of magpies, and will even scheme with her male partner to dupe the unsuspecting host parents. Dad will swoop in and stage an attack on the magpie host parents, and, while the hosts are fending off the attack, mom will swoop in and quickly lay her eggs. Indeed, many brood parasitic moms have evolved extremely fast egg-laying strategies to enable them to deposit-and-run. While the process of egg laying in most non-parasite bird species is approximately twenty minutes, a brown-headed cowbird can lay an egg in forty-one seconds, a shiny cowbird can lay hers in approximately seven seconds, and bronzed cowbirds can lay their eggs in an astonishing five seconds.

The egg has reached the nest, but will it be discovered and rejected? Brood parasites have evolved sophisticated means by which to fool even the most discerning host moms. For example, females of several cuckoo species display complex systems of egg-mimicry to achieve success. They undertake several observational excursions of host nests prior to selecting the one that they will parasitize. It's a maternal reconnaissance mission, an attempt to match the chromatic composition of the eggs that are already in the nest. Egg appearance is an important characteristic that may make the difference between acceptance or rejection of a specific host egg—and so matching the hues precisely is a helpful skill. Mothers of these specialist species have evolved complex physiological mechanisms to be able to fine tune the color of their eggs to more exactly match those of the host. And the match is quite precise, since birds have better color vision than humans do; they can see colors in our visible spectra as well as in the UV spectrum. Parasite eggs with the highest levels of color overlap with host eggs are less likely to be rejected by host parents, offering a greater level of survival for these offspring.

In addition to their sophisticated color vision, most bird species—and especially those involved in brood parasitic relationships—have mechanisms to detect even minute changes to egg patterning and texture. The common cuckoo is a generalist parasite which has evolved a remarkable ability to create eggs of a wide variety of colors, textures, and patterns depending on the host being parasitized at any given time. Aspects of patterning include size of markings, number of markings, diversity in size and density, and contrast and dispersion of markings. There are virtually innumerable combinations of these characteristics that create species-specific egg morphology, and parasite moms like those of the common cuckoo have evolved the ability to mimic eggs to match the specific individuals that they are parasitizing.

Timing of egg laying is another important characteristic of consideration for brood parasites. Parasite moms will ultimately aim to lay their eggs around the same time as those of their hosts to capitalize on the opportunity to quietly add a few extra eggs to the collection without the host mom noticing. Hosts that are not in the process of laying are more likely to identify foreign eggs as such and dispose of them accordingly. Indeed, this is the case with the host fairy-wren and the parasitic Horsfield's bronze cuckoo. If the parasite eggs are laid before the fairy-wrens begin laying, the eggs are discarded. However, if the cuckoo holds off laying until the fairy-wren's own period of laying begins, their eggs are no longer detected. If the cuckoo lays her eggs after the main laying period of the wren, she will simply desert the nest and none of the eggs will successfully develop. Therefore, it pays for the cuckoo to be vigilant about the exact time frame of egg laying in its host.

In the perfect scenario, a brood parasite mom would lay her eggs during the laying period of the host and have an incubation time which sees her eggs hatching a few days earlier than those of the host mom. This way, the parasitic chicks will already be larger and more competitive than the host chicks from the very beginning. In many parasite-host systems, this is exactly what is observed. Just the opposite scenario would be the case if a parasite mom timed the laying of her eggs such that her chicks hatched *last*—the chicks would have a lesser chance at effectively competing for food due to their small size and competitive ability. In addition, if a parasite mom laid her eggs much later than a host mom, after most of the egg incubating by the host mom had already been done, the parasite eggs may not receive enough incubation time to develop appropriately and may not hatch.

If the egg laying of parasite and host moms is out of synch, and a parasite mom has missed an opportunity to parasitize a preferred nest, she may engage in a behavior called nest-farming. Nest farming moms will destroy entire egg clutches or broods of host moms to reset the process of egg laying, so that she can now have a greater chance of successfully parasitizing the nest.

All the previous discussion paints a pretty bleak picture for host moms, who have the unfortunate reality of having to defend their nests against the reproductive exploits of another species. But the host moms aren't without a few tricks up their sleeves. Maternal defenses evolved to mitigate the damages incurred from three main stages of the parasitism process: prior to the laying of foreign eggs, after the laying of eggs but before hatching, and after hatching.

Among the most effective strategies any mother has for minimizing nest parasitism is to prevent it before it begins. These frontline defenses occur prior to egg laying of any kind, either host or parasite. The location of the nest is a prime example of a frontline defense. Many host species build their nests away from perching points of parasites so that they are less likely to be found. Camouflage is key. If parasite moms cannot find the nest, they cannot parasitize it. Host parents themselves exhibit a high degree of camouflage; many have evolved to have dull coloration to prevent being found by parasites. In fact, host moms are less likely to sing characteristic songs or make mating calls to minimize exposure.

To make matters even more complicated, some parasite species have evolved to appear similar in size, shape, color, and behavior to *predators* of their hosts. Such is the case with the gray and the common cuckoo, the reed warbler, and the Eurasian sparrowhawk. The sparrow hawk is a predator of the reed warbler, which in turn is a host to the gray morph cuckoo. In mimicking the appearance and behavior of the predatory sparrowhawk, the parasite cuckoo elicits anti-predator defense behaviors in the warbler host, which allows the cuckoo to find the inconspicuous warbler nests. Similarly, two species of cowbirds—the shiny-headed cowbird and the brown cowbird—make a series of short, noisy flights to flush their hosts.

Host species can "fight back" against such techniques by building their nests within the vicinity of third party species that the parasites would rather avoid. For example, some hosts will construct their nests beneath the nests of large predatory birds like hawks or eagles or close to colonies of stinging insects. Yellow warblers will often build their nests close to nests of red-winged blackbirds, the latter of which are extremely aggressive toward their cowbird parasites. In this way, host species build their homes in environments that are naturally inhospitable to their enemies.

Structure of the host nest may also play a role in deterring potential parasites. Some host species build their nests in small tree holes or cavities. Cavity nesters enjoy relative safety in their homes from a variety of environmental factors including weather and predation, and they may also incur a lower level of nest parasitism. It's difficult for parasite moms to get their eggs into the small cavity nests, and even if they manage to do so, their chicks may not be able to get out so easily. For example, cuckoo chicks are quite larger—often larger than the host chicks—and they may not be able to fit through the exit from the cavity, ensuring their demise. African weaver birds build large, complex nests that often have entry tunnels up to twelve inches in length. This long-entry tunnel makes it very difficult for parasite moms to enter.

Great spotted cuckoos prefer to lay their eggs in large magpie nests because large nests are indicative of high territory quality and parental care. Magpies have evolved to build smaller nests when the incidence of parasitism is high, as a defense against being targeted. In this way, the magpies are taking measures to directly deceive the parasites. Song thrush moms have developed another way to remain reproductively successful: by creating nests that are deep with steep sides. This unique nest architecture renders the parasites unable to remove host eggs from the nest. Once the eggs hatch, the song thrushes preferentially feed their own chicks, and the cuckoo parasite chicks starve to death.

Several other aspects of nesting and behavior can come into play at the frontline of defense against brood parasitism. Host species may live in dense aggregations to overwhelm the parasite need for nests, and to benefit from a whole collection of nest defending neighbors. Mobbing behaviors by host species are often quite effective at decreasing the ease with which a parasite can lay eggs in a host nest. Many host populations live in close proximity and synchronize their egg laying to reap the benefits of parasite swamping and group defense. If all host species lay eggs during a specific time, there is a greater chance that parasitic birds won't be able to add eggs to all of the host nests.

Sexual system also plays a role in the success of nest parasitism. Monogamous species are less likely to be parasitized because mom or dad is almost always constantly guarding the nest. In polygynous species where one male mates with many females, nests are apt to be less guarded, leading to the potential for increased parasitism.

Many host species manage to successfully evade parasitism through the frontline defenses, but many others do not. Having foreign eggs in one's nest is a conundrum faced by various host species, and the fact that many parasite species have evolved the ability to create extremely effective mimic eggs does not make matters easier. There are a few different evolutionary directions that a host mom can take once the egg has landed in her nest. The first is to evolve a means by which to distinguish her eggs from impostors. As mentioned earlier, birds see colors in several spectra including UV, so the ability of a parasite mom to create closely mimicking eggs can be matched or rivaled by a host mom's ability to discern even very slight differences between eggs. Second, a host mom's eggs could become so varied in appearance between clutches that while eggs of one specific clutch all resemble one another, they look nothing like the eggs from another clutch. She'd be able to recognize her own eggs, but increasing the variety would make it that much more difficult for a parasitic mom to accurately mimic the eggs of any given clutch.

For some host moms, it is not a matter preventing the laying of parasite eggs, but about attempting to minimize the level of damage that occurs. The shiny cowbird is a parasite that punctures the eggs of its hosts. Punctured eggs fail to develop, and so the successful cowbird brood parasite not only gets its offspring raised by the host parent, it also kills the offspring of the host. Many host species will attempt to drive away incoming parasitic species by flying directly at them, striking, biting, or knocking them off their flight path in a behavior termed "mobbing." While mobbing behavior is intended to deter the laying of parasitic eggs, for species like the chalk browed mockingbird, which battles against the shiny cowbird, it is a defense of their own eggs. The mobbing behavior undertaken by the mockingbirds prohibits the cowbirds from getting a chance to puncture their eggs. Yellow warblers do not engage in mobbing behavior but instead remain steadfast on their eggs when approached by a parasitic cowbird. By remaining on top of their own eggs and allowing the cowbird females to approach and lay additional eggs into their own nests, which will roll down the side and join the warbler's eggs when she moves, the warblers don't avoid the parasitic eggs, but they do prevent the potentially fatal damage that the cowbird moms would otherwise inflict on their host eggs.

If the host species can tell the difference between their own eggs and those of a parasite, they can eject them or bury them within their nest. For example, yellow warblers who know that their nest has been infiltrated will often create a new nest lining over the top of the foreign eggs and initiate a new clutch in the superimposed nest. It's not entirely clear if the warbler moms can tell which egg is which; they may simply understand that eggs other than their own were added to the nest. The byproduct of this is that the warblers will often inadvertently bury some of their own eggs while burying the parasite eggs.

The final stage at which a host mom can discern whether she has been parasitized is once the eggs have hatched. There are several species in which moms can identify their own chicks over the parasite chicks and they simply avoid feeding the latter. Superb fairy-wren moms can discern their own chicks from those of their nest parasite, the common cuckoo. These diligent moms allow the parasite chicks to starve to death, which usually happens within a few days of hatching. However, the ability to tell the difference is fairly rare. If a host mom is unable to tell the difference between her own chicks and parasite chicks, the parasite chicks stand a good chance of not only surviving, but thriving. The stage is set for competition between host and parasite chicks, and the latter are generally more equipped for the win. Many non-evicting parasite chicks have evolved rigorous begging behaviors and in many host bird species (and many species, period) it is the loudest and most demanding chicks who obtain the most nutrition. In addition, there are many parasitic species where chicks are larger and much more aggressive than the host chicks. Parasite chicks in these scenarios have a clear advantage over host chicks in that their mouth gapes are more visible and higher up in the nest than those of their host competitors. In addition to being louder, parasitic shiny cowbird chicks have distinctly different calls than the chicks of their fairy-wren hosts. The cowbird calls contain repeated syllables rather than the one simple chirp exhibited by the fairy-wren chicks. These multi-syllabic calls are termed "tremulous" calls, and they function akin to a repeated form of the host chick call. The parents respond to this clamoring by ramping up provisioning to the cowbird chicks. They do this because the tremulous calls exploit a common provisioning rule in birds: repeated signaling is associated with greater demand and thus results in a greater transfer of food. The cowbirds are the ultimate tricksters, begging for more and getting it.

In cases like these, where there is the potential for intense competition between parasite and host chicks, there is likely to be strong selection pressure on the ways in which chicks grow and develop. It is advantageous for host-species to evolve growth strategies that allow them to better evade their parasites, including a faster rate of chick growth, an earlier development of chick flying ability, or more intense chick-begging behavior. Indeed, a meta-analysis (a large statistical study encompassing the results of several smaller ones) of 134 North American passerine species and their brood parasitic partners showed that high levels of parasitism are associated with a faster rate of chick growth, a lower mass at fledging, and shorter nesting periods overall. The competition scenarios that play out between host and parasite chicks provide a plethora of material by which biologists can observe the effects of evolution in action. After a host species evolves a new means by which to avoid a parasite, the parasite then evolves a counter-means to trump it. This is ultimately what biologists mean by an evolutionary arms race.

Although some non-evicting brood parasites will compete with their hosts, evicting parasites spell the certain demise of the entire host species brood. Egg-laying parasite moms can expel the host mom's eggs from the nest or puncture the eggs without getting rid of them. "Eviction" can also happen after hatching when parasite chicks repeatedly stab host chicks with specialized bill hooks evolved specifically for that purpose. Such is the case with striped cuckoo parasites and their rufous and white wren hosts. The cuckoo chicks violently kill all the wren chicks soon after hatching with their sharp mandibular and maxillary hooks. Evolutionarily speaking, evicting parasitic species have no need to appear like the young of their host species or to evolve more rigorous begging behavior since all of the host species are going to be killed anyway.

Despite the aforementioned selection pressures and strategies for brood parasites, there remain many other examples of brood parasitism that do not fit neatly in any kind of definitive box. There are host species that simply provide parental care to non-mimetic, non-evicting parasites. It's thought that this could represent a kind of "safety" strategy for the host parents, where they aren't quite sure enough if the foreign chicks are in fact foreign. It could also represent a lag in the evolutionary arms race, where a new parasite strategy has evolved but a host's counter-strategy has not yet evolved.

In any case, the entire existence of brood parasitism across the bird world represents an astonishing maternal strategy that is most certainly not what one would expect in a traditional context of motherhood. Humans tend to place our own values and judgements on the behaviors of the rest of the members of the animal kingdom in a rather flippant way. However, brood parasitism represents a sound and successful evolutionary strategy. Who are we to judge?

4

COOPERATIVE CREEDS
AND COMMUNAL CLIQUES

Many females in the animal kingdom have a massive battle in front of them to even become mothers in the first place. It isn't generally a battle with fertility; it's a battle for a social position that affords a female the chance at reproduction. Think *Handmaid's Tale* meets *The Origin of Species*. Whereas most human females make up their minds to become pregnant outside of the immediate influence of other females in their social group, this is *not* the case for cooperatively breeding mammalian species, where complex social dynamics play a major role in the probability that any given female will become a mother. While it's no surprise that most males in the mammalian world achieve reproductive success by using their physical traits (tusks, feathers, teeth, ornaments), it's less obvious that many females achieve it through effective use of social strategies. It's *Mean Girls* in the animal kingdom.

In cooperatively breeding mammals such as meerkats, naked mole rats, and several social primates, there is a dominant breeding couple. Other members of the group are subordinate. They do not sexually reproduce; instead, they help the dominant couple to raise their offspring. Helpers take on a number of roles including foraging, nest- and home-building, defense of the group from predators or competitors, babysitting, and allonursing, or wet nursing. In many cases, the female subordinate group members are related to the dominant female; they could be her sisters, daughters, aunts, or cousins. Males are more likely to disperse from their natal groups, meaning that there are likely to be more unrelated males in a group.

Subordinate helper females remain celibate and do not attempt to mate despite being sexually mature and having ready access to unrelated males in the group. Biologists believe that these females retain the possibility to become mothers at some point in the future if a suitable situation presents itself; however, for the most part, they do not take any action toward this end. But why is this? The benefits of cooperative breeding are widespread in insects, birds, mammals, and primates, evidence that it is an evolutionarily favored strategy. Subordinate females can still maintain an adequate level of biological fitness if they are genetically related to the dominant ones—and they are likely to experience violence, maltreatment, or even infanticide if they become sexually active, making the "choice" to help more like deciding on the lesser of two evils.

In meerkats, subordinate females are dutiful toward dominant ones, grooming them, suckling their babies, and generally doing their best to remain in good favor. There can be serious backlash from dominant females toward submissive ones for any kind of misbehavior—including attempts at mating. Dominant females, who undergo a secondary growth spurt once they obtain a breeding position and are therefore of considerably larger size than submissives, will kill the pups of their submissive counterparts on the rare occasion that the submissives become pregnant. Low-ranking females are often the sexually mature daughters of the dominant, meaning that the alpha female often kills her own grand-offspring. Dominants may also be physically abusive toward submissive females; they bite, scratch, fight—they intimidate the submissive to prevent mating and pregnancy. Older (and thus larger) submissive females represent a greater threat to the dominant, so they ramp up their subservient behavior toward the dominant, especially when she is pregnant. The dominant will evict other females in the weeks leading up to her birth to reduce the likelihood that the submissives will kill her pups while she is in a vulnerable state. If a dominant female suspects any kind of aggression from a submissive during her pregnancy, she will be expelled from the group. This eviction can be intensely stressful. Females lose weight, have increased levels of stress hormones, and experience spontaneous miscarriages if they are pregnant at the time of eviction. Eviction can be seen as "encouragement" from the alpha female for the evicted female to permanently remove herself from the group and start a new group elsewhere. It's a classic case of the alpha female taking advantage of her social status to get rid of her closest competition. Whether it's the evil stepsisters in *Cinderella* or the evil dominant female meerkat, such themes are present in many species across both the mammalian world and the entire animal kingdom.

To avoid being evicted, submissives try to maintain a good working relationship with the dominant—especially when they are older and represent a greater threat. Part of this investment includes providing nutrition to the newborns of the dominant in the form of wet-nursing, or allonursing. What's unique about this behavior is that there has been no direct demonstrated physical benefit of this extra milk to either the pups or to their dominant mother. The pups do not gain more weight from the additional milk supply. The dominant female does not have a shorter interval between births or cut down at all in her own nursing of her offspring. This points toward the potential for an entirely social function of the allonursing behavior in meerkats. In fact, subordinate females are even more likely to allonurse the offspring of the dominant if they have recently experienced an eviction from the nest. It's perhaps a way these females demonstrate to the dominant full acceptance of their lower social status.

But how does a certain female become the queen of the tribe in the first place? In meerkats, unlike in many other mammals, dominance is not determined through genetics. A pup born to a dominant mom has no advantage when it comes to whether she will become dominant herself. This makes sense, as a dominant female meerkat has an average of seventy pups during her reproductive tenure—far more than other mammals with comparable lifespans. Such a huge number of offspring means that one is no more likely than any other to be "special." This astoundingly high number of pups born to the alpha female is only possible because she has an average of fifteen helpers (and usually more) at her beck and call at each step of the way. However, only approximately 17% of meerkat females ever have a chance at becoming the dominant female, and there are only two means by which this can happen. They can become the alpha in their own natal group by displacing the current alpha through intense aggression or they can leave their natal group and begin a different society as their leader. The latter usually happens when a subdominant female is evicted by a very aggressive alpha and then leaves the group permanently along with closely related females who will benefit from being in a new regime via their closer genetic relatedness to the new alpha.

Since the alpha female bears the most of the group's litters, it's more likely that her successor will be one of her offspring rather than a pup born to a submissive mom; however, the large number of offspring suggests that some other characteristics give certain females a competitive advantage. Individual pups with higher body weight are certainly favored as better competitors overall. But context is important too. If an alpha female dies due to some unforeseen event, her position will likely go to the female with the next highest body weight and age who can out-compete others at that exact moment. Younger females could become larger than a new alpha female at some later time in their lives, but the queen will be decided by the circumstances only at the time when they are required. The greatest competitors are those with the highest (and most similar) body weights. It's the ultimate girl fight, and the losing challenger will likely be severely punished by the new alpha as she secures her dominant position. Punishment can include intense aggression and, ultimately, eviction.

Another mammalian system in which such division of labor is seen is the naked mole rats. Naked mole rats live mostly underground in elaborate tunnels, and rarely come above ground for any reason other than to obtain food, defend the colony, or found a new colony (the last of which is rare). Unlike the subordinate female meerkats, who are fertile but do not reproduce, all subordinate female mole rats are infertile. A specific chemical cue contained in the urine of the single dominant female keeps all other females from reaching reproductive maturity. In this way, she maintains a total monopoly over reproduction with a relatively low level of dissention in the ranks.

The queen does all the reproducing in the colony with the help of between one and three male partners. All other subordinate males are also infertile but retain the ability to become fertile if the correct social parameters present themselves. Naked mole rat societies are the only mammalian societies known as being totally eusocial—with a clear, pronounced, and irreversible division of labor among the infertile members. There are defenders, resource gatherers, maintenance laborers, waste managers, and of course babysitters. However, if the queen dies, there is an immense battle within the underground colony between subordinate females determined to take her place. Such intense fighting has major impacts on the functioning of the naked mole rat societies, and in extreme cases may result in their dissolution. Eventually, one female physically dominates all others to become the new queen. A period of approximately six months ensues as she settles down and physiologically matures. She will then begin to secrete the "anti-maturing" compound in her urine to prevent the other females from becoming reproductive.

Although it seems at first glance to be quite remarkable that the queen uses a chemical to keep her potential female competitors infertile, this kind of control is common in many mammalian species. Historically, steroids have been thought to originate and have use within each individual's body; however, when it comes to reproduction, there are vast biological gains to be made through using one's own hormones to manipulate the reproductive physiology of a potential partner (or competitor, in the case of the naked mole rats). This phenomenon is most frequently studied in mice. A chemical called estradiol (E2) and other compounds in a male's urine have strong impacts on a female's reproductive tract. Females that are already pregnant can experience spontaneous miscarriage if males deliberately direct droplets of urine at a close enough proximity to them. Conversely, female mice that are not yet sexually mature will experience a fast-tracked version of puberty as a result of the same steroid. The E2 present in the male's urine is absorbed through the sensitive tissues on the female's nose or directly through her skin, and it makes its way to her ovaries and uterus, where it accumulates and works directly on her reproductive physiology. This phenomenon is well-documented across several mammalian families including mice, voles, hamsters, possums, pigs, and even primates. It seems that in many cases, females are not in direct control over many aspects of their reproductive physiology due to the impacts of the males around them.

The effects of urine-based steroids are also observed between females, as the naked mole rat example shows. The estrus cycles of females living in groups without males become either elongated or totally suppressed. Estradiol from the urine has been shown time and time again to be the culprit for such changes to the reproductive function in these females (a phenomenon also documented in our own species). When a male is introduced to the group, his pheromones both jumpstart the female cycling process as well as induce synchronicity within it. This describes yet another fact of chemosignaling in mammals that is directly related to a female's reproductive status. While at first glimpse it may appear quite phenomenal that the queen mole rat is able to suppress sexual maturation in her female colleagues, it is a common practice in several mammalian groups for diverse reasons. However, it is fair to say that the immense physiological influence of *only one* female in a mammalian population is quite remarkable.

From this point of view, it seems like being the mole rat queen is a fairly cushy and powerful position. However, when one begins to break down what this entails, the job becomes far less desirable. The mole rat queen is little more than a baby-popping, newborn-suckling machine. Reproduction is continuous in this species, as opposed to having a specified breeding season, which means that the queen can become pregnant year-round. Gestation time is between sixty-six and seventy-six days, and the average litter size is twelve. (The spine of the queen elongates during each pregnancy, so she becomes much longer instead of wider—a phenomenon that makes sense for tunnel-dwellers.) Once the pups are born, they suckle exclusively from mom for two weeks, and then they begin to incorporate other foods in addition to suckling for a few more weeks. After this short period of "rest," the queen will become pregnant again with another litter. The long lifespan of the naked mole rat means that over the course of her reproductive life, a queen can have more than 900 pups.

In addition to considering the sheer number of offspring that the mole rat queen is responsible for producing, it's important to acknowledge the immense lactation effort that this requires. Biologists have calculated that, for the first two weeks postpartum when the female is exclusively providing milk for her pups, she has to produce over 50% of her own body weight in milk each and every day. The milk of naked mole rats is very low in actual nutrition and very high in water content, which is meaningful in a few ways. First, newborn mole rats are at risk for water loss through evaporation (since they have no fur and have a higher surface area to volume ratio than adults) and they do not develop the physiological ability to concentrate their urine until around two weeks old. This means that they really need a lot of water for their first few weeks to survive. But what about mom? To produce more than half of her own body weight in milk (even diluted milk), she's got to be ingesting a massive amount of nutrition and water herself. Indeed, she has many helpers bringing her an endless supply of water-rich tubers and other high-moisture foods so she can feed the hungry newborns. A queen mole rat consumes up to ten times more than non-reproductive females. (No kidding.) Although on the surface being the queen appears substantially cool—being the only female with sexual access to males, having all the food one could possibly eat brought directly to you, and having many helpers to attend the babies once they come along—overall it seems to me like this is one completely overworked mom.

In cooperatively breeding hyenas, social dominance is determined along matrilines: if mom was dominant, her female pups will be dominant as well. In hyena populations, both high- and low-ranking females do reproduce (unlike meerkats and mole rats), but the opportunities for high-ranking pups are much greater. Low-ranking hyena females and their pups have a lower access to quality food, dens, and other resources. Conversely, high-ranking females and their pups have first access to any kill made by the group, as well as to prime den sites, and so the cycle of dominance continues. The circle of privilege is not simply at the individual level either. In large groups of social primates with several matrilines present, females of high-ranking matrilines, who are usually genetically related at some level, can lend support to members of their privileged circles, such that the powerful stay in power and the subordinates have a much more difficult time getting ahead. It goes without saying that we also see this in our human social systems.

This kind of cycle of support and violence toward specific females can have profound impacts on overall group function. In societies where dominant females have tight control over submissives—engaging in effective dominance and infanticide where necessary—group size generally remains small. However, when dominant females intimidate submissives but do not successfully prevent them from reproducing, group sizes can become much larger. These larger groups, seen in animals such as savannah baboons and spotted hyenas, can have multiple factions that each have their dominant mating pair but also lower ranking individuals that are also close kin, including their own offspring and siblings. Close families will stick together and support each other, essentially creating mini-alliances within a greater population. When life is good and resources are plentiful, one can imagine that such societies function in a peaceful manner. However, when resources are scarce or conditions are otherwise suboptimal (catastrophic weather events, death of a dominant), family members are more likely to actively discriminate against non-kin.

It has been observed in spotted hyenas that dominant offspring need not be the genetic sire of the dominant female. If a dominant female adopts a pup that she hasn't birthed, a rare but natural occurrence, that pup acquires the benefits of its adopted mother's rank. Conversely, if a hyena pup is adopted by a lower-ranking mother, the pup itself becomes low-ranking as well, regardless of genetic background. These kinds of scenarios allow us to understand the functioning of maternal support and how maternal support shapes many aspects of the offspring's future existence, even outside of the influence of maternal genetics and physiological mechanisms during pregnancy. It turns out that mom's social and behavioral support as pups grow and mature is also an important factor in determining their level of social standing as adults. This is a hugely important point that speaks to the notion of nature versus nurture in the context of motherhood in hyenas.

Due to the distinct advantages of being female, there is a high risk for intense competition between siblings within a litter, with especially aggressive competition directed toward female infants. Hyena pups are born ready for battle, with highly developed jaws and sharp teeth. They are essentially mini killing machines. If a pup is male, it would be more advantageous for him to kill his sisters so that he secures more resources for himself. He "knows" that his sister is guaranteed to be of a higher social dominance to him. Sister pups would want to kill each other to avoid competition for resources and to eventually become the highest-ranking female in the group. In all cases, it's best for any pup to kill a sister—but pups may not discern the sex of their siblings. Male and female hyenas have evolved to look identical to one another—especially as juveniles—and potential for siblicide may be a reason. Females have pseudopenises, which are extensions of clitoral tissue that look identical to a male's penis. They behave like penises too. Females can get erections—which are often used in dominance displays—and they urinate through their pseudopenis. In addition, females have externally fused labia that resemble scrotum, and so it is literally impossible to tell the sex of a hyena from a mere observation of their external anatomy.

There is another distinct advantage to being a female with a penis. Male hyenas cannot have sex with females unless the females retract their pseudopenises within their bodies (rather like turning a sock inside out) to create a cavity where a male can insert his penis. Unless a female retracts her pseudopenis, there will be zero chance for copulation. In the mammalian world, sexual coercion is extremely common, but due to this remarkable physical structure, it's impossible for rape or sexual coercion to occur in hyenas. Males cannot have sex with females unless the females are interested parties. Consider that the hyena male's inability to rape females is concurrent with an entirely female-dominated society. Coincidence? Hardly.

There are also considerable downsides to having these penises. In addition to using their penises for social dominance, the female hyenas also need to give birth through them. You can rest assured that the process is not pretty. But more on that later.

The New World monkeys of Central and South America, tamarins and marmosets, are the main groups of cooperatively breeding primates. These are collectively known as callitrichids, and they inhabit many areas of Central and South America and share many characteristics that render them good candidates for cooperative breeding. Females always give birth to twins, which are dizygotic but that share a placenta. Adult callitrichids are small, so a twin birth represents a major investment for adult females. In fact, the twins are heavy at birth and represent between 16 and 20% of the mother's total body mass. In addition, female callitrichids can get pregnant again within a few days of giving birth. This means that reproductive females in callitrichid groups have an abundance of work to do—so much, in fact, that they rely on the help of others. The reproductive systems in these primates rely on the provisioning of offspring by non-parental helpers. In fact, the *cooperative breeding hypothesis* was coined to describe the scenario where successful sexual reproduction is not possible without the extra help of non-breeding adults. Non-reproducing females and all males play a substantial role in infant care in both tamarin and marmoset populations. From the first day of birth, alloparents and caregivers take the twins (because it's difficult for new mothers to carry two newborns) and only bring them to the mother to feed. Callitrichid infants are carried non-stop for the first two weeks of their lives; they are still carried to a large extent for the subsequent ten weeks after that. Carrying them generally results in weight loss of the helping individuals, so the job of "infant carrying" is usually relegated to larger helpers. Smaller helpers (even infants from five months of age engage in helping behavior) are charged with tasks like nest maintenance or food collection. Similar size-related task patterns are seen in other cooperatively breeding mammals: large naked mole rat helpers contribute more to territory defense, whereas smaller helpers are tasked with infant care, and small meerkat helpers provision young and help with tunnel digging, whereas larger ones babysit and guard the nest.

It was once believed that some callitrichid species were monogamous, making them the only monogamous primates. However, more detailed studies in recent decades have revealed that there is a great deal of fluidity in the sexual systems of these primates. Some groups are polygamous, with multiple males mating with one dominant female, some are polyandrous, with two reproductively active females, and some groups defy a specific classification as members may engage in sexual activities with individuals from another population. For the most part, the breeding systems of these primates can be described as cooperative, where most females do not themselves engage in sexual activities, or even ovulate. Reproductive suppression is common in callitrichid populations, but it's rarely absolute.

In some marmoset populations, a second female within the group becomes reproductive. This generally happens when the group has become sufficiently large that there are more than enough helpers to assist with the dominant female's offspring and other group tasks. In other words, the pressure of achieving the group's tasks is lessened when there are more hands around to do the work. Such a scenario requires "sneaky sex" between a non-breeding female and either one of the group's males or a male outside the group, and it produces another set of twins for the group to look after. Subordinate females that undertake this sneaky breeding tend to be extremely protective of their newborn twins, rarely letting any other individuals near during the first few weeks—quite unlike the *laissez faire* attitude of dominant females, who hand off their babies to caregivers often from the very first day. This extreme protectiveness is in response to the threat of violence or even infanticide from the dominant female who may kill the subordinate's offspring or disallow their provision. In a truly remarkable set of circumstances, the offspring of the dominant female have access to the milk of the subordinate female even before the subordinate female's own offspring. It's a very unfortunate scenario to think of a mother having to starve her own offspring while she feeds the offspring of the reproductive dominant. Complicating things further, the offspring of the subordinate are likely to be somehow related to the dominant female. Females are the philopatric (i.e. they do not disperse from their natal home) sex in these marmoset groups, and so a dominant and subordinate could very easily be a mother and daughter or a pair of sisters, making their offspring directly related.

In some marmoset populations, the eldest daughter of the dominant female begins to ovulate and engage in copulations if an additional, unrelated male is added to the mix. The quality of the relationship between the specific mother-daughter dyads in question is an important factor in determining the outcome. If daughters are very intimidated by and submissive to their mothers, chances are they will not become ovulatory or engage in sexual activities with a novel male. However, if a daughter is more independent from her mother, the presence of a novel male will induce sexual behavior and ovulation. In the latter scenario, the mother and her daughter may work together to suppress ovulation and sexual behavior in other females to keep the reproductive success within their own family. This shows that the lack of ovulatory activity in subordinate female primates could be the result of both behavioral intimidation by the dominant female or it could be a physiological response to inbreeding avoidance (i.e. females can remain non-ovulating if the only reproductive male around is their father). The presence of a novel, sexually mature male may be all that is required for some females to begin ovulating—and it's therefore best for the dominant female's biological fitness if the second reproducing female is one of her own daughters.

In tamarin populations, the only time a subordinate female can successfully rear offspring is when the dominant female doesn't have a litter. Similar to the marmosets, if the subordinate female engaged in that "sneaky sex" and managed to give birth to twins, their best chance at survival would be when the dominant female didn't have a competing set of twins. In this way, others could provide help to the subordinate female that they couldn't otherwise provide if they were busy with the offspring of the dominant. It's mostly a matter of luck—perhaps a form of biological gambling on the part of the subordinate female. Since there is no anestrous phase for callitrichids (i.e. there is no reproductive down time between births—dominant females usually get pregnant again right away and gestate for four to six months), there is always a risk for subordinate females in terms of being sexually active or reproductive. However, if the timing is right, a subordinate female stands to dramatically increase her biological fitness.

Humans are among the many animals considered to be cooperative breeders, though our large brains, emotions, and the need to fulfill our own biological destinies complicate matters somewhat. There are several hypotheses to explain cooperative breeding in primates (humans included), including an argument for inclusive fitness. Genetic relatedness between females in a group means they are more likely to have a vested interest in the offspring of the breeding female (perhaps their own nieces, nephews, cousins etc.). Also, the *parenting experience hypothesis* explains that being a subordinate helper may be advantageous for younger females who need to learn the basics of parenting prior to becoming reproductively active themselves. The *group augmentation hypothesis* suggests that there are substantial benefits to be gained from living in a larger group—things like predator avoidance. Since females are philopatric and will remain with the group anyway, it's advantageous for their direct survival if that group is larger. Individual females can help to keep it large, or make it larger, by assisting with the offspring created by the dominant female. The *pay-to-stay hypothesis* suggests that subordinate females look after the offspring of dominants as a means by which to pay the dominants for the right to remain within the group. In this way, subordinate females may be biding their time by remaining in the group until a breeding opportunity arises.

Although some aspect of each of these hypotheses may explain *some* aspects of cooperative breeding in callitrichids, it's important to remember that we are talking about complex and intelligent animals here, and there isn't likely to be one specific "right" answer for this kind of behavior. Primates have large brains capable of distinct personalities and emotions; some female callitrichids may simply *enjoy* helping, and they may even prefer to undertake certain infant-care tasks over others. In a similar fashion to human day-care providers, there are some females that are just happier doing it than others. Although there are biological benefits to be gained from being the dominant breeding female, the chances of attaining this role are slim-to-none for most other females in the group. It's certainly possible that some individuals *understand* that they are unlikely to become dominant, and that if they engage in sex (and reproduction) anyway, their offspring has an attenuated chance at best. It's possible that these females make the decision not to cause this kind of "trouble" and to remain in line with the expectations of the group. We would be silly to ignore the important role of individual personality in the reproductive decisions made by female callitrichids.

As discussed above, cooperative breeding can take place in two ways: singular cooperative breeding, in which only the most dominant female is "allowed" to reproduce, and plural cooperative breeding, in which more than one female is permitted to breed. Another kind of breeding system exists in several mammalian groups in which many females are reproductively active and they raise young communally, helping with each other's offspring. This is quite distinct from cooperative breeding, and there are both costs and benefits to its practice. Communal care could favor the evolution of unequal investment by the mothers involved, or cheating: mothers may take advantage of the shared help to have more offspring than others. There could also be increased exposure to parasites or bacterial disease, or increased competition between mothers. Additionally, larger groups can be easier targets for predators. But there are plenty of benefits of communal care as well, including an increased level of total energy to spend on oneself, foraging or otherwise, increased immunological factors based on exposure to many different milks, lower cost of child care, shared information with other moms, defense against predators, and a decreased costs of thermoregulation.

One of the only primate groups to practice communal breeding are the lemurs. Black and white ruffed lemurs are small-bodied primates that live in small groups on the island of Madagascar. Reproductive females appear to have a choice between communal or solitary rearing of their offspring. Some females synchronously brood offspring in crèches (nursery groups) and act communally to groom, protect, and feed them all. While there are definite risks to leaving one's offspring in the care of allomothers, there are also benefits to be gained from the greater level of variable care. Indeed, ruffed lemur moms that participate in communal breeding have an overall higher level of reproductive success than those that do not. Communal breeding *can* work if the females taking part are not related; however, it makes the most sense from an inclusive biological fitness point of view if they are. Most of the communal primate and mammal breeding groups include females and their close kin, as there are clear advantages to helping ensure the biological success of close relatives, and the costs of helping related offspring are lower than the costs of helping those that are not.

The biology of ruffed lemurs makes them excellent candidates for communal breeding. Females only come into estrus for roughly twenty-four to seventy-two hours per year. This means that the breeding of all females is highly synchronous: all females will have young pups to care for at approximately the same time. There are numerous benefits to "buddying up" with another or a few females that are in a similar stage of motherhood, including allomaternal care, infant provisioning, and protection from predators. Lemurs also exhibit what is termed a "boom bust" reproductive system, which may be a function of living on a small island with unpredictable and severe environmental conditions. In some years females may have large litters of offspring, and in others there may be zero reproductive success. This means that following a time of "bust," mothers will be extremely motivated to maximize the number of offspring that they can successfully rear when times are good.

Since all moms in a synchronous group have similar requirements with respect to newborn defense and care, and they all produce highly nutritious milk that their newborns need, there is enormous potential to experience the benefits of all working together for a common good. This is especially true when it comes at a time when moms are highly motivated to ensure reproductive success. Mothers who rear offspring in groups tend to have much higher newborn survival rates, as well as improved maternal energetics, because they can take turns looking after *themselves* while their newborns are at the communal den with helpers.

Despite the clear benefits to be gained, there are always females in ruffed lemur groups that do not participate in communal breeding. These moms give birth to and feed their offspring away from the communal dens, on their own. Individual personality differences may contribute to a female's decision about whether to breed with others, as active exclusion of any female from a group has not been observed. In some lemur societies, when a female's closest relatives die, she begins to breed solitarily, eventually setting up a new group comprised of herself and her grown daughters.

As mentioned before, females are the philopatric sex in ruffed lemurs, so there is a good chance that a female has some relatives in any given group. There are cases where immigration and/or extenuating circumstances might mean that a female is without biological kin, but that doesn't necessarily stop her from communally breeding with non-relatives. There appears to be a great deal of fluidity with which lemur moms choose their path for a given litter. Females engaging in communal breeding may be doing so with non-relatives, and solitary breeders may be doing so despite having full sisters in the group.

Another communally breeding primate from Madagascar is the gray mouse lemur—the smallest primate species known. Mouse lemurs are generally solitary, but mothers are nevertheless plural breeders. They forage alone during the night, but during the day, individuals form nursing groups with either one or two other females, usually closely related kin. Mothers in these groups will communally feed and defend all the offspring within them. All female mouse lemurs give birth to one to three pups at a time, and they only move their *own* pups between different daytime resting sites. Such behaviors indicate that these moms can discern their own offspring; however, they continue to provide allomaternal care to other pups in the daytime crèches by indiscriminately grooming and feeding all offspring in the nest. This is an important point to consider—the continuing cognizance that one is providing care to offspring that are biologically unrelated, and that some reciprocal assistance is likely. If a mom is killed by a predator, chances are another mom in her crèche will adopt her pups and continue to provide complete care to them. In this way, communally nesting females have a kind of "insurance" that their pups will be taken care of if they die.

It's true that communal breeding and care of offspring is generally uncommon in primates other than humans. Communal breeding is much more common in mammals such as carnivores, rodents, and bats, and the reasons relate to the specific ecology of the species in question. For example, female lions pool all their offspring into crèches, and non-offspring nursing is ubiquitous. Lions live in stable social units called prides, which are fluid groups of between one and eighteen adult females, between one and nine adult males, and their dependent offspring. All females in a pride are relatives, and they were either born into their pride or founded it with members of their own natal group. Adult males form a coalition that "guards" their females, and paternity of all cubs in the pride is shared among them. Approximately every two years, the pride is overthrown by a new coalition of adult males, who will take over the group of females as their own. All dependent cubs at the time of the overthrowing are killed by the new coalition, which serves to bring all females back into estrus at the same time, and they will all give birth to new baby cubs approximately three and a half months later.

Female lions remain sexually active throughout the year. They have four teats and give birth to one to four cubs at a time. However, cubs are kept in hidden dens by their own mothers for their first six to eight weeks of life. These dens are not communal; mothers remain with their own biological cubs only, feeding them exclusively. It's thought that prior to reaching eight weeks of age, cubs would be unable to defend their own mom's teats against the eager mouths of foreign cubs, so moms keep them separated until they have learned to do this. After such time, all females bring their cubs together to form the crèche and begin communally raising the cubs.

Although non-offspring nursing is very common in carnivores like lions, there is no evidence of *direct* reciprocity within groups. Females give away up to 30% of their milk, and much of this occurs while they are resting and sleeping at the den. The only time that mothers are away from their offspring is when they go off to hunt. When they come back for periods of rest, they are subjected to frequent suckling attempts from both their own cubs and other cubs in the crèche. It would likely take a much greater effort for a female to discern between her own and foreign cubs and then constantly remove the latter while she is trying to rest. In this case, moms save energy by indiscriminately allowing cubs to nurse from her teats while she is resting. Her own cubs are old enough to fight for their own share, both from her own teats and those of other mothers in the crèche, so mothers simply allow the foreign milk consumption to take place without really paying much attention to whether all mothers are equally reciprocating. In addition, an increased amount of suckling activity on any mammalian female's teats works to stimulate further milk production, which is good news for all cubs in a specific crèche, assuming that moms can obtain all of the nutrition that they require to produce the extra milk. Females that produce small litters nurse offspring as much as females with large litters, which indicates that some females are potentially putting in a greater effort than others for the good of the entire group. This seems a little counterintuitive—why provide milk to so many offspring when you've only just had one of your own and you don't necessarily need to rely on the milk supply of other moms?—until you consider the main reason why female lions form communal groups in the first place.

Cub mortality is an unfortunate reality of lion life. Infanticide by male intruders accounts for approximately 27% of all infant death. Mothers stand a far greater chance of effectively protecting their offspring if they band together in a group, so they band together to protect their offspring from invading males. Although a female may "lose" some of her milk supply to non-offspring cubs while she is resting or sleeping in the crèche, it's well worth it to have the support of all the females in the group when defending their collective offspring. In addition, a female is at least partially related to the other females in her group, so there is also some support for increasing biological fitness through kin selection. A lone female would likely have a low rate of success in rearing any cubs all on her own.

A low success rate for solitarily breeding is also likely to be the case for mothers of another mammalian group—bats—although the reasons for communal roosting in these mammals are quite different. Female evening bats share common roosting areas, and despite being the philopatric sex, they do not necessarily form groups with kin versus non-kin. As with the nocturnal mouse lemurs, female bats go on nightly foraging trips and come back to communal roosting sites where many of them have young pups. Bats have specific maternity roosts, where females go to give birth and spend their first few postpartum hours until the pups can fly. Mothers then make use of communal nursing roosts, where many females with dependent offspring may interact, rest, and provide care to each other's pups. These communal roosts are large, with up to several hundred breeding females and their pups sharing a common space.

As the only flying mammals, bat mothers find themselves in a unique situation in the animal kingdom when it comes to lactation. At the same time that they provide milk to their offspring, they also need to fly to forage sites and find food for themselves. A heavy load of milk has a substantially negative impact on a female's ability to fly, and so it makes a lot of sense for bat-moms to get rid of excess milk prior to heading out. This is known as the *milk dumping hypothesis*, and it accurately describes how female bats can keep their milk production at a high level while still efficiently foraging for themselves. At peak lactation, a female evening bat can produce just over half of her own body weight in milk per day. This is a substantial amount of extra mass for an individual that needs to fly to find food. What this conundrum means for females who must forage and fly is that if they have older pups who aren't drinking as much milk (and are hunting for themselves), they must lose their extra milk-weight prior to setting out on a foraging excursion.

Just how is a mother supposed to lose her extra milk weight? By providing this nutritious food resource to another pup in the communal den. In this way, allonursing in bats seems like a win-win situation, with bat moms using their excess milk to feed other pups in the roost—something that ultimately functions to make the entire colony larger and stronger. Females belonging to larger roosts with healthy individuals benefit from information and predator protection, so ultimately it is beneficial for them to help the group by donating their excess milk. Moms with older pups also have the advantage of a continued milk supply to provide to their older pups just in case they have a bad day foraging for themselves. The act of suckling by foreign pups keeps a bat mom's milk supply strong.

Protection from predators is a major factor for any small mammalian species, and it is not something that should be considered lightly. By roosting in large groups, females not only protect themselves and their offspring, but they also have the potential to learn from other members of the group. Things like foraging sites, predator awareness, and resource gathering efficiency can all be learned by observing successful group members. This scenario is termed the *information sharing hypothesis*. By indirectly yet significantly contributing to the success of the group via communal lactation, a female is doing her part to keep the group strong.

Thermoregulation is an additional aspect for careful consideration by female bats when selecting a communal roosting area. Bats have a very high surface-area-to-volume ratio, and they can lose heat through their skin quickly. With their small bodies and membranous wings, staying warm while resting is a prime concern. Therefore, it's critically important for females to roost where they can maintain stable body temperatures for both themselves and their pups. The density of females in a specific area contributes to thermoregulation; more bodies means more heat to be shared among them. In fact, females roosting together can raise the temperature in their dens by up to 44 degrees Fahrenheit.

Because bats are so sensitive to heat loss, they have evolved the ability to enter a state of torpor, or physiological slowdown that is characterized by decreases in both metabolic rate and overall body temperature. Torpor is a kind of mini-hibernation state that can happen on a much shorter time scale than a full-blown hibernation—and it happens in a small number of mammal and bird species that have a requirement to conserve energy and/or water during times of resource shortage. Female bats of varying species select warmer roosts to avoid torpor, which carries greater costs for lactating moms than it does for other adults. However, torpor does happen to bat moms if they get cold enough—and it slows down milk production, which ultimately results in a slower rate of infant growth. If a mom is pregnant while she is in torpor, it serves to lengthen the gestation time of her pups. Depending on the individual circumstances, this could have a negative impact on the ultimate growth and survival of her pups.

Indeed, some bat-moms in Australian species select *extremely* hot spots for roosting and communal nursing. Microclimate is exceedingly important for these species, and specific aspects of the local substrate may become important factors for moms to consider. For example, in southeast Australia, female east-coast free-tailed bats select mangrove forests, where tree roots exist in a gnarly mass in ocean shallows, for their communal roosts. Not only do mangroves have an abundance of small crevices and hiding spots, but temperatures within them remain relatively consistent and they can be humid, which prevents dehydration in juvenile bats at night when their mothers are out foraging.

Female brown bats utilize the microclimates of rock crevices for their maternity and lactation roosts, which are deeper and more thermally stable and warmer than non-maternity ones. Roosts with very specific characteristics are utilized by females at different times during their reproductive cycles, which reflects an intricate and well established system of optimal body temperatures. Rock faces are surprisingly complex in terms of their structure and formation of roost sites and microclimates, which makes them excellent for the specific requirements of female bats. In addition to being more stable in temperature, lactation roosts are insulating, so they retain heat for the young bats who do not leave during the night. They are also cool enough during the day to allow for shallow torpor to take place in lactating moms. The extremely high costs of lactation require that mothers "catch up" on energy saving through use of torpor during the day; however, lactating females enter a shallower torpor than non-lactating females, indicating a kind of medium level of unresponsiveness in the mothers. This makes a lot of sense, seeing as lactating moms need to remain vigilant for their pups; however, they do require rest periods to maintain milk production. Once pups have weaned, females utilize cooler roosts and are able to enter a deeper state of torpor to make up for the energy lost during lactation. Pregnant females may also select cooler roosting sites during phases of poor resource availability, and, in that state, extend their pregnancies until resources become availabile again.

Communal offspring care is also common in many rodent species. Female mice give birth to litters of offspring, which they either choose to rear alone or with one or several other females. When female mice give birth at approximately the same time, they often pool their litters. However, in doing so, they lose their ability to tell the pups apart. They therefore go on to take care of all offspring equally—grooming, nursing, and protecting them.

This non-discrimination of offspring occurs even if the contributing females' litter sizes are different. In other words, a female may contribute two pups to the nest—or she may contribute eight. Once the offspring are pooled together, she no longer knows which two were hers, and now she's providing communal care to ten offspring.

The mother who had a litter of eight only has to provide communal care to two extras in order to receive communal care for her eight offspring. This is clearly an unfair scenario for the first mother. However, equal sharing of communal duties is still observed. How could such a system evolve? What would be the benefit of each female losing her ability to identify her own offspring? It may be a phenomenon that is not directly beneficial to the individual mouse-moms involved. Female house mice may be the victims of a genetic hijack.

House mice and many other rodent species are polygynous, meaning that one male mates with most of the females in his immediate vicinity. As a result, there is little question that he is the father of the pups birthed by those females. For the father, the best-case scenario is if all his pups are well cared for and thrive through to sexual maturity. The best way for him to accomplish this is to have all the mothers involved look after all the offspring together, without conflict. In this way, there is a distinct conflict between what dad wants—lots of care for all the offspring—and what mom wants—to make sure she isn't providing too much care to her own and other offspring so that she still retains a chance to reproduce in the future by maintaining her own health. Something called *imprinted genes* play a role in the resolution of this kind of conflict between the sexes.

Imprinted genes (also discussed in the next chapter on mammalian development) are a special class of genes that work in parent-of-origin specific ways. Only one allele of the gene is expressed based on the sex of the parent from which it came. In other words, a *paternally expressed* imprinted gene is active in the offspring according to what was inherited from dad. The maternal copy of the gene remains *silent*. The notion of gene imprinting and conflict between parents is also an important aspect during embryonic development—where dad would prefer that mom give all of her energy and nutrients to the baby whereas mom would prefer to hold back and save some energy for herself and for her future reproductive efforts. I discuss this in much greater detail in the gestation and pregnancy chapter. In this example, the paternally expressed gene is active at a much later phase of development. Indeed, most imprinted genes have some kind of a regulatory function and their expression is limited to very specific times during development. During the lactational phase, the maternal odor cues associated with infant recognition are turned off, which means that mom cannot tell her own pups from any other. The mechanism of "turning off" mom's ability to recognize her own offspring comes directly from dad. Seeing as all of the pups have the same dad but different moms, this scenario represents a unique opportunity for dad to make sure that all pups get indiscriminate care. After the pups are weaned and inbreeding avoidance has the potential to interfere with a female's reproduction, the paternally suppressed gene may cease functioning and allow for discrimination once again.

That's not to say that communal mouse nests are happy, wonderful, conflict-free places to be. Maternal mice can be extremely violent toward each other's infants if they are born at a different time of the cycle and therefore do not have congruent needs. Timing is a crucial part of the puzzle when it comes to surviving in a communally breeding species. When pups are born to communally breeding females at approximately the same time, they look so similar that when pooled it isn't visually obvious whose pups belong to whom. However, if another mother joins an established group and has a newborn litter, it is immediately apparent to the other mothers that some of the pups are the wrong size. These pups are then generally killed by the other mothers of the group. It sounds harsh, but keep in mind that not all mice are infanticidal. In fact, many aren't. However, the instance of infanticide is high enough that counterstrategies such as synchronous breeding have evolved in response. The threat of infanticide may be one of the reasons behind the evolution of solitary breeding in some individuals of various mouse species, and many female mice retain a degree of plasticity when it comes to their reproductive lives. For example, female striped mice can do one of three different things when it comes to reproduction: they can breed communally with other females and provide allomaternal care to pups, they can give birth away from a shared nest (presumably to avoid the potential for infanticide) and then return to the nest once the pups are old enough to remain by their side, or they can give birth away from the nest and remain away from the nest with their pups and never return to the group. Such a solitary existence is usually selected by females that are large and heavy, as they have the energy reserves and the competitive ability to make it on their own with their pups.

Other small rodents like prairie dogs have different factors to consider when it comes to communal breeding—things like immense predation pressure. Prairie dogs provide a good deal of allonursing support to pups that do not belong to them; the instance of non-offspring nursing in these ground-dwelling mammals is a whopping 25%. It's thought that in these animals, mothers are providing a service that is akin to misdirected parental care, since mothers are not able to tell the difference between their own pups and other similar aged pups in the colony after an initial five-and-a-half-week period of keeping them hidden underground. Once offspring emerge from their underground burrows, they start to mix with other pups, and juveniles from one litter may spend the night with juveniles of a completely different litter as soon as four days after they emerge from their natal burrows. Pups continue to nurse above ground from assorted mothers for part of their diet until they are fully weaned around eight months later. In prairie dogs females are the philopatric sex, and so they remain within the natal colony whereas males leave the colony at sexual maturity. For these moms, it seems that the reduced predation pressure that the group experiences is the major advantage, and the fact that moms provide nutrition to non-offspring pups reflects a somewhat small cost of colonial living. Like many group-living prey animals, having an abundance of individuals to "watch your back" at any given time is a massive advantage. This, combined with the potential for relatedness between pups and reproductive females, means that the advantages of group living outweigh the costs of misdirected maternal care.

The previous discussion of examples from several mammalian groups shows that there is a vast variety of reasons why females may choose to rear their offspring together as opposed to alone. One final example from the rodent world is the degus, which is a plurally breeding social rodent endemic to central Chile. Mothers provide care to all offspring in the group, regardless of relatedness, and this suite of maternal substitutes can serve to provide offspring with a buffer to stress. Although that initially reads as being quite complicated, any time that more than one mom provides care to an infant, that infant has the potential to benefit from the interaction. Each mom will have something slightly different to offer, and in the case of the European badgers the offering can take the form of a buffer against maternal stress.

When mothers are really stressed, the levels of stress hormones in their milk is accordingly high, and these glucocorticoids have negative impacts on infant physiology and behavior. In addition, stressed mothers may fail to perform maternal tasks, and things like licking and grooming—which are very important for normal physiological and emotional development—are often skipped altogether. However, when many mothers are contributing care and/or milk to an infant, the stress factors from one source are diluted. This is most certainly beneficial for the infant that would otherwise have a steady supply of stress hormones and potentially a lack of grooming and licking from its own mother. This kind of potential for immune factors or endocrine signals from many moms to have positive impacts on infant physiology makes a lot of sense; however, it hasn't yet been widely studied. Plural breeders have the potential to provide and derive benefits from a wide variety of sources; however, as with anything, those benefits need to outweigh the costs of providing care to unrelated infants or dealing with the fallout of a crafty cheater.

5

BUNS IN MAMMALIAN OVENS

When humans think of gestation, we invariably draw on what we know happens in our own species: fertilization, nine months, and then along comes baby. We have ample resources to let us know what kind of symptoms to expect and when, how big the baby is at any given time, and the developmental timing of its various physiological systems along the way. This process isn't the same for all mammalian species. While there are many important parallels, there is vast diversity in the process of development from embryo to neonate within the mammalian group. Mammals display an immense array of shapes, sizes, and textures, and they can live on land, in the water, in the air, or any combination thereof. What works well for the gestational process in one group will naturally reflect its specific ecological requirements, and so it stands to reason that we see a wide variety of strategies across the mammalian order.

There are three main groups of mammals: the Prototheuria, the Metatheria, and the Eutheria. The names translate respectively to "first beasts," "behind true beasts," and "true beasts," although those classifications are misleading. Yes, the Prototheurians were the first to diverge from a common mammalian ancestor, but the others aren't so cleanly classified, and it's generally dangerous when a name implies that there is some kind of evolutionary race to become the "most derived" or "most evolved" animal. The fact of the matter is that this could not be further from the truth. Evolution does not, and never will, imply that animals are working toward being some kind of "best" animal. We all exist in our own current evolutionary end-point states.

The Prototheria are monotremes, such as the echidnas (spiny anteaters) and platypuses of Australia and New Guinea. Their developmental pathways are a combination of reptilian, avian, and mammalian characteristics. Monotreme mothers are physiologically specialized to produce both eggs and milk, and they have a single urogenital opening for both reproductive and waste products. After a very short period of internal gestation (twenty-one days for the short-beaked echidna and fifteen to twenty-one days for the platypus), females lay leathery eggs, each containing a developing fetus.

Platypus moms generally lay two eggs, sometimes three, and echidna moms usually lay just one, but all monotremes create underground burrows for the protection of the mother and her offspring. The burrows created by platypuses can reach over twenty-six feet in length, and they usually end in a spherical nesting chamber. Moms collect wet vegetation and keep it in the chamber to create a humid environment for the eggs and nestlings. Platypus eggs are laid in the humid underground chambers, while echidna eggs are laid directly into an abdominal pouch and are subsequently carried by the mother.

Platypus moms can also incubate their eggs by holding them closely between their tail and abdomen, although this prevents them from moving around extensively, so it's less common.

Eggs are incubated for a short period (10 to 12 days in platypus, 10 to 10 and a half days in the echidna) and they hatch at an early stage of development that is akin to a human giving birth during the first trimester of pregnancy. At this point, all monotreme moms and pups begin a very long lactational phase (114 to 145 days in platypus and 160 to 210 days in echidna). Neonates are moved by their mothers from incubation burrows to nursing burrows (although echidna pups can still spend time in their mom's pouch), and they are constantly tended to by their mothers. Milk is dilute when it first begins to flow, which is critically important, as the digestive systems of the tiny newborns are not able to handle much else at such an early stage of development. The milk becomes higher in fat and protein as the newborns become older and their digestive systems begin to mature. It's believed that the original functions of lactation in monotremes had nothing to do with nutrition. Instead, the dilute formula served to prevent eggs from drying out, and its antimicrobial compounds worked to safeguard the offspring from infection as they hatched. Lactation has secondarily evolved to become a source of nutrition in these mammals, and there is an abundance of antimicrobial compounds in the milk from the very beginning.

Monotreme moms are the only mammalian moms that do not have nipples. Milk secretion occurs through areolae on abdominal skin patches, and these areolae are covered with fur or hair. Otherwise, the system of ducts and alveoli in the mammary glands is similar to that of other mammals. Tiny newborns suck on the areolae on the patches and obtain nutrient-rich milk as well as a plethora of antibacterial and antimicrobial compounds. The need for immune factors in monotreme milk is tremendously high, as the neonates have a very poorly developed immune system when they hatch from the eggs. In addition to the lack of a physiological means by which to fight infection, the underground burrows undoubtedly expose hatchlings to a high level of bacteria and other pathogens. As the infants grow and become stronger within the underground burrows, there is less of a requirement for moms to visit as often. Mom will eventually stop provisioning her babies altogether, encouraging them to become more and more independent.

The next group of mammals is the Metatherians, which includes marsupials such as kangaroos and wallabies. Like the monotreme hatchlings, marsupials have newborns that are termed "highly altricial," meaning that they are extremely underdeveloped at birth—approximately the stage of a human fetus at around just eight to ten weeks' gestation. When marsupials give birth, they sit on the ground with their tails in front of them and between their legs, which facilitates the ensuing journey of their tiny newborns. Birth happens after a very short gestational period (26 and a half days in wallabies, 31 to 36 days in kangaroos), and the neonates propel themselves into the world through their mother's urogenital opening using their tiny forearms. They then crawl, unassisted, upward along the mother's abdomen to reach the entrance of her pouch. Once at the entrance of the pouch, they abruptly switch directions and crawl down into the pouch toward the teats. Although the marsupial neonates are not well-developed at the time of birth, their sense of smell matures early, and they use olfactory cues to guide them to the mother's pouch opening and then her teats. Just around the time of birth, mothers lick the furry pathway between their pouch and their genital opening to keep the pathway moist and clean; it is likely that this creates a sensory roadmap for their new offspring.

Once safely affixed to a teat within the mother's pouch, neonate marsupials remain there for the next several months. Upon latching, milk flow begins. Mothers initially supply small volumes of dilute milk to their newborns because their digestive systems remain fairly undeveloped (another similarity to the monotremes, although newborn marsupial milk is less dilute than that supplied to newly hatched monotremes). Marsupial milk is initially comprised of simple sugars with almost no protein or fat content. The lungs of the neonates are only partially developed, and so the bulk of respiration occurs through direct gas exchange with the skin. The newborns are also ectothermic, so they obtain their body heat from their mother. In short, "newborn" kangaroos and wallabies are more like "partial born," as they still have a lot of foundational developing to do from the safety of mom's pouch.

As with the monotreme newborns, the immune system of marsupial newborns is extremely underdeveloped. It isn't until approximately one month after they enter the pouch that they are able to mount any kind of immune response. So how do they manage to make their way from the urogenital opening of their moms, which is hardly a sanitary spot, to the pouch without being attacked by microbes or pathogens? Indeed, over seventy-two types of bacteria have been identified from the urogenital openings of female tammar wallabies, and a further fifty from the anus, which is only a fraction of an inch away from their point of origin. In addition, there is also a great deal of bacterial diversity within the pouch itself, so newborns must certainly require some mechanism by which to stave off infection during the first few days of life. These tiny pups are born with an outer epidermal layer called a periderm, which serves a protective function and lasts for approximately a week. In addition, several immune-boosting compounds are shared through mom's milk, and the pouch itself secretes some anti-bacterial and anti-microbial compounds. Far from being a simple "pocket," the pouch is an interactive and complex physiological system.

Survival of such undeveloped newborns wouldn't be possible without many specialized adaptations of the marsupial pouch. First, it keeps mom and baby in direct contact with each other, which allows for a constant neonatal temperature, making up for the lack of endothermic capability in newborns. Mothers also control the level of humidity within the pouch, which facilitates gas exchange through the newborn's skin. The level of gas exchange varies by species; in tammar wallabies, it's a whopping 33% of the neonate's oxygen supply at first, and it tapers to around 14% at six days postpartum. The pouch itself is also a direct physical barrier that affords protection to the developing newborns from the elements and other negative aspects of the outside world including predators and infection. The pouch allows moms to exploit ecological niches and environments that would not be available to animals whose young are carried on their backs because of the increased protection during jumping and gliding. In addition to being well-protected, marsupial newborns are also inconspicuous, which affords further protection from outside threat.

In wallabies, the first stage of teat-latching lasts for approximately 100 days, after which the neonate sucks less often but remains confined to its cozy pocket. The first forays out of the pouch happen after around 200 days, although at the beginning, these sojourns are just for short periods of time. Milk composition changes a lot during this phase of development, with a shift away from sugars to a greater concentration on proteins and fats. Unlike in humans and other primates, lactation in marsupials is multifaceted: females tailor-make milk depending on the specific developmental stage of the young. Each of the mammary glands of tammar wallabies functions independently of the others, so that females can simultaneously feed a neonate milk that is of completely different composition to that consumed by an older, out-of-the-pouch juvenile.

The highly specialized and unusual developmental strategies of monotremes and marsupials may have evolved to provide parents with the option to terminate a pregnancy if environmental conditions quickly change and become unfavorable. It makes evolutionary sense for a "kill switch"-type mechanism to be in place *early* in the pregnancy rather than later so that investment in doomed young can be quickly and easily terminated. In the Australian outback, where habitats are unpredictable and extreme, selection for a developmental strategy where not much is lost could be strong. Monotremes and marsupials have a very rapid birth, and a low fetal cost in terms of time and energy. In addition, mothers retain the ability to resorb reproductive material following a terminated pregnancy, so the actual costs of losing an infant are comparatively small for these animals. Moreover, if a pregnancy is terminated or a neonate is lost, a marsupial can almost always make up for the lost infant by producing another one immediately.

The final group of mammals is likely the one that you are most familiar with: Eutherian mammals, which includes everything from cats and dogs to giraffes, zebras, and elephants, to dolphins and whales, to monkeys and apes, and, of course, ourselves. Developmentally, we are all characterized by a major reliance on an exclusively mammalian maternal organ: the placenta.

During the Eutherian ovulatory cycle, a structure called the corpus luteum is derived from the ovarian follicle. Once the egg bursts free from its follicle, the corpus luteum is left behind. It's this structure that is responsible for establishing and maintaining pregnancy in all Eutherian mammals. Progesterone is secreted by the corpus luteum, which functions to thicken the lining of the uterus in anticipation of embryonic implantation. If the egg isn't fertilized, the corpus luteum will break down after a few weeks and either be resorbed by the body or will exit the body as a menstrual period.

Progesterone is also involved with additional cellular factors in preventing the activation of the maternal immune system against the fetus. The fetus expresses paternal or "non-self" antigens, which for all intents and purposes should otherwise be attacked and disposed of by the maternal immune defense system. (When you think about it, that's what a pregnancy really is: An invasion and takeover of the female body by a foreign entity.) However, the body must circumvent its own immune defenses for females to be reproductively successful, and progesterone has a large role to play in this regard. Females have evolved two main ways to keep up with progesterone production. The first is to maintain the life of the corpus luteum, and the second is to have another organ that takes over progesterone secretion. In most mammals this organ is the placenta.

Eutherian pregnancy is vastly different from pregnancy in either monotremes or marsupials, and the placenta is one of the factors that sets them apart. Monotremes and marsupials have something called *intra-cyclic gestation*, which means that the maternal reproductive cycle is not affected by the presence of a fetus; their fetuses don't stay in the female's uterus long enough to have a significant impact on the ovarian cycle. The same is not true for Eutherian mammals, who have *trans-cyclic gestation*. The maternal hormonal cycle *is* affected by the presence of a fetus, and maternal physiology is distinctly altered in a variety of ways. One characteristic of the pregnancy is the increased and sustained production of progesterone, a key factor in the evolution of trans-cyclic gestation.

The placenta is the most variable mammalian organ, despite serving a similar function in all lineages. It plays a massive role in providing nutrition and oxygen to the developing fetus, and it also carries away the waste products that the fetus creates. These are both critical and fundamental functions in all mammals despite the huge variation we see in their overall physical structures. There are three major placental characteristics that are used to classify them into distinct types. These are its shape (including the number and position of gas and nutritive exchange areas), its level of interdigitation (or the spatial arrangement of maternal and fetal tissues with respect to each other, a key aspect of the surface area available for gas and nutrient exchange), and the level of invasiveness (the number of maternal tissue layers that separate the maternal and fetal bodies). In highly invasive placentas, fetal tissues erode maternal tissues to gain more direct access to the mother's bloodstream.

Keeping these morphological characteristics in mind, there are three basic kinds of placentas in the mammalian class. The first is epitheliochorial placentas, which are non-invasive because the fetal tissues do not breach the tissues of the mother's uterus. There is a distinct separation between mother and baby despite the baby growing inside the mother's womb. The mother's three basic tissue layers (the epithelium, the connective tissue, and the endothelium) act as a barrier between the fetal tissues and the maternal blood-stream. Animals with epitheliochorial placentas include cattle, sheep, pigs, dolphins, and whales.

The second kind is endotheliochorial placentas, which are more invasive than the epithelial ones. In this case, there is an increased level of contact between the fetal and maternal tissues. Carnivores and elephants have this type of placenta, which is in direct contact with the endothelial wall of maternal blood vessels. It is this very limited placenta-to-fetus interface that may be one of the reasons for the exceedingly long gestation time experienced by elephants.

Elephants have an exceptional reproductive cycle. Mothers carry their developing offspring for twenty-two months. After a calf is born, they experience a lactational anestrous (no ovulation during the phase of lactation) of approximately two years more, and they may have an interbirth interval of between three and seven years. Although this is most certainly a substantial amount of time, elephants are also extremely long-lived mammals, and females remain reproductively active until almost the end of their sixty- to sixty-five-year lifespans. Generally, larger animals give birth to larger babies and have longer pregnancies. However, elephants stay pregnant far longer than is predicted by this general trend, and they have longer gestation times than several whale species that are much larger.

Elephants maintain a steady stream of progesterone during their long pregnancies not from their placentas, but from two to twelve accessory corpus lutea that remain active in progesterone secretion for the entire twenty-two-month duration. This is quite different from most other mammalian systems that show a degeneration of the corpus luteum after embryonic implantation when the placenta takes over progesterone secretion. Dugongs and manatees—aquatic mammals that are close relatives to the elephant—also have accessory corpus lutea. In addition, elephant moms also get an extra surge of progesterone from the gonads of their own developing offspring late in their gestation.

Although the elephant placenta does not secrete progesterone, it still provides the bulk of nutrition and oxygen to the developing calf. Remarkably, the placental connection between mother and calf in elephants is quite small, as mentioned earlier. Something called the *hilus* is the only part of the placenta that connects the mother's endometrium to the fetus, and this is only three-quarters to one-and-a-half inches wide. This kind of connection only permits very slow fetal development. And, although baby elephants develop quite slowly, they are born in a very precocial, or capable, state. These newborns make immediate use of their trunks and recognize their position in the complex matriarchal social structure. Because of all of this, it takes a long time to cook. The advanced development of the baby elephant's central nervous system prior to birth, combined with the small placental-fetus interface, contribute to its lengthy gestation.

The last type of placenta is the hemochorial, where there is direct contact between the mother's bloodstream and the fetus.

Fetal tissues are bathed in maternal blood, facilitating nutrient and resource transfer. This kind of placenta is observed in rodents, primates, and of course humans. The facilitated transfer between mother and fetus helps animals with hemochorial placentas grow faster, and promises a greater level of maternal involvement.

An obvious question that comes to mind when describing all the diversity in placental structure is *why*? Why is it necessary for placentas to be so variable, and how have they evolved to do similar things in such different ways? These are questions that have been puzzling biologists for decades. Since the relationship between mother and fetus is so multifaceted, there is no easy explanation for the existence of placental diversity. In the case of the hemochorial placentas, where mother and baby are physiologically joined via the placenta, the opportunity arises for there to be a kind of competition between them for resources. Indeed, a kind of evolutionary war exists within a mom's womb between the offspring, who would like to maximize its nutrient uptake, and the mother, who would like to preserve some for herself and her future offspring. The scenario gets much more complicated from there. In addition to the conflict between mother and child with respect to resource allocation in utero, there is also a phenomenon called gene imprinting (discussed briefly in the previous chapter), which carries the additional input of the baby's father. Imprinted genes are expressed in the fetus from only *one* parent. What does this mean? It means that only one copy of these specific genes will be expressed in the fetus—either the copy from dad *or* the copy from mom (not both, as in most genetic characteristics).

Paternal genes are generally in conflict with maternal genes against the mother in her own uterus to divert more resources (glucose, lipids, amino acids) toward the developing fetus. When imprinted genes are paternally based, the mother loses her ability to control the overall consequences of resource uptake by the developing baby. While the paternal gene says "give more!," the maternal gene says "I've given enough!" It is this conflict that underlies the evolution of genomic imprinting in hemochorial placental mammals, and it echoes the conflict that was already present between mother and offspring.

While paternally imprinted genes promote a level of placental and fetal growth that goes beyond the optimum level of the best interests of the mother, the maternal-imprinted genes work to combat this trend and to restrict both fetal and placental growth. This situation becomes even more complicated when one considers that mom can be pregnant with a litter of pups (e.g. carnivores, rodents) who are either full siblings or half-siblings to each other (she may have received sperm from more than one male). Paternal-imprinted genes from different dads can then be competing with each other and against the interests of the mother. It's a chaotic scenario to say the least. Over 100 imprinted genes have been identified in mice and humans, and these are predominantly expressed in early stages of development, at the embryonic, fetal, and neonatal stages. Biologists can selectively knock out specific genes from either mom or dad, allowing them to decipher the specific roles of each in fetal and placental development, but this is a painstaking process that is in its infancy.

Although the understanding of gene imprinting and parent off-spring conflict within the fetus and the placenta is exceedingly complex, the very idea of a kind of war taking place within a female's womb is utterly extraordinary. The best part is that mothers-to-be are likely blissfully unaware of the entire process, that their own physiological processes are working hard to ensure success against the enemies in her own uterus. Sounds like something out of a science fiction movie.

6

A PREGNANT PAUSE

You may operate under the false impression that once a sperm has met with an egg and created an embryo, the process cannot be paused and restarted. However, the physiology of many mammalian and primate females allows them to shift their fertilized embryos to a state of diapause—or arrested development—if certain environmental characteristics incite them to do so. Embryonic diapause is a very common occurrence in the invertebrate world, where encysted (dormant) forms can exist in hidden crevices for many seasons until appropriate environmental conditions wake them from their prolonged rest.

In mammals, embryos can exist in a state of suspended animation for many months. It's thought that the ability of mammalian moms to essentially store fertilized embryos originally evolved to maximize reproductive success, especially in areas where resources are patchy or seasonal. The first documentation of arrested development of mammalian embryos came from observations on the roe deer. As early as 1843, biologists understood that there was some kind of delay of embryonic implantation that lasted between four and five months; however, the mechanisms and evolutionary origin of such a phenomenon were poorly understood. The developmental delay prolongs the gestation of the roe deer to a total of nine months, and offspring are delivered in the optimal season for their continued existence.

Over 150 years later, we understand a great deal more about the process of diapause, but there remains mystery surrounding its existence and mechanisms of control. Arrested development of the embryo at the blastocyst (early developmental) stage is observed in over 130 mammalian species, and it's hypothesized that the ancestral condition of all mammals is to exhibit some form of embryonic diapause. This is because it has been observed in representative species of all major mammalian groups, indicating that the ability to encyst embryos has always been an important factor in normal development through evolutionary time.

In harsh climates, where further development of offspring could be risky, it makes a lot of sense for mothers to be able to put the development of their offspring on hold until conditions become more suitable for continued development. Entry into diapause can be controlled by day length, presence of suckling offspring, or availability of nutrition. For example, female brown bears vary the birth date of their cubs in response to their own accumulated fat stores. Those in superior condition give birth earlier than those in poorer health. This is a commonality of many northern mammalian moms, who retain the ability to birth their newborns during the most favorable time of year for finding food, being mobile, and avoiding predators.

Many of these females cannot predict or control when they may meet a viable male, and so some mechanism of decoupling embryonic implantation from copulation is clearly advantageous.

The blastocyst can be thought of as the ideal phase of development during which it is possible to pause. In all species that exhibit diapause, it's always at this exact same stage of development. Once egg and sperm have united and created an embryo, development occurs until the blastocyst stage. Unless some specific uterine signals are present, development stops here. Despite being in a lowered state of both protein and carbohydrate synthesis, the dormant blastocyst remains healthy and at the ready to receive signals to proceed. The uterus and embryo must have some kind of critical signaling communication in order for the blastocyst to resume development and eventually implant into the uterus.

In embryonic diapause, the uterus either sends a signal for continued development, or it doesn't. If uterine signals are absent, development stops. When the signals are present, development continues. When diapaused embryos reanimate, it happens very quickly; both DNA and RNA synthesis resume and cell division and metabolism continue within four to sixteen hours after the resumption of development. In a fascinating study that has revolutionized how we view the evolutionary origins of diapause (Grazyna Ptak et al. 2012), researchers were able to transplant sheep embryos to a diapausing mouse uterus and keep them there for a specific interval before re-transplanting them back to the sheep where they developed into normal lambs. Sheep do not normally exhibit embryonic diapause, but this experiment confirmed that their embryos were unharmed by the process.

These results are groundbreaking because they confirm that diapause doesn't harm embryos of mammals that do not normally exhibit it, and this makes it more likely that its existence is widespread and perhaps even hidden. It's possible that sheep retain their ancestral ability to encyst embryos; they simply do not normally need to use it because environmental conditions (especially those in domesticated environments where food and protection are ample) are adequate for reproduction to continue. The same group of researchers who transplanted the sheep embryos has also transplanted rabbit and cow blastocysts into mouse uteri, again with successful results. This outcome speaks to the ancestral condition of all mammals as being one that includes a diapausing stage.

The animal systems that have been examined the most when it comes to embryonic diapause are rodents, carnivores, and marsupials. In each of these groups, the diapause is similar but also different—and the specific molecular factors that occur at each stage are far from being clearly understood. There are two kinds of diapause, however, and as we learn more about the specific molecular controls, it may turn out that these terms describe different aspects of the same process. For now, they remain useful descriptors of the process in different animals.

First, obligately diapausing mothers experience a period of embryonic inactivity each time a pregnancy occurs. Regardless of environmental conditions, there is a programmed pause in embryonic development for every embryo, and the duration of the pause varies according to the species. Photoperiod (day length) is often associated with obligate diapause, as this is a specific and defined cue that impacts many aspects of animal physiology. Facultatively diapausing mothers, on the other hand, show arrested development that is related to specific environmental or physiological conditions. Facultative moms also exhibit fine-tuned control over the length of the suspended state. In other words, they retain the ability to pause a pregnancy depending on specific extenuating factors that they are facing.

For example, if there is a resource shortage, an outbreak of predators, or an environmental catastrophe, facultatively diapausing females can hold off on continuing their pregnancies until the outlook is a little brighter. Mouse and rat mothers are examples of such species. Females ovulate immediately following birth, and they then copulate, creating another litter of embryos. The newly minted embryos reach the blastocyst stage by the fourth day following copulation, when they enter a state of diapause. During this phase, the embryos space themselves evenly in the uterine crypts of the mother, where they remain until they receive a signal to become reanimated.

As opposed to the rodents described above, carnivore moms exhibit obligate diapause. For example, pinnipeds, including seals and sea lions, are widely dispersed in their aquatic habitats for much of the year, although they have very well-defined breeding seasons. Mothers give birth in large rookeries at approximately the same time as each other despite variation in the timing of copulation and conception in the wild. Evolutionarily, it's tremendously advantageous for all pinniped offspring to be born in synch and to have them reach independence at a similar time. This overwhelms the predator populations (since the predators cannot possibly kill and eat all the pups that are available), which increases the chances of survival for any one pup. Predation is an unfortunate and normal part of existence for pinnipeds, but embryonic diapause helps mothers maximize the chances of survival for their offspring.

Other well-studied carnivore systems of embryonic diapause are skunk and mink. Skunks mate in the autumn, and embryos enter a 200-day diapause before reanimating. Mom skunks give birth in the late spring when their offspring have the best chance at survival. Coincidence? Not quite. Mink breed in early- to mid-March, and embryos undergo just a brief diapause prior to being birthed just over a month later. Unlike the mouse blastocysts that space out evenly during diapause, the mink blastocysts remain in a cluster near the anterior portion of the mother's uterus. Once they have been reanimated, the blastocysts space themselves out more evenly for further development. For both the skunk and the mink, it's the advent of the vernal equinox that provides the initial cues to the female's physiology to bring the dormant embryos back to life and resume normal development. For another carnivore, the red fox, reproductive delays associated with embryonic diapause are only observed over parts of the species range, which indicates flexibility in the ability of mothers to partake in it. It would be a very interesting experiment to transplant females between diapausing and non-diapausing populations to see what happens as a result.

In addition to the mammalian mothers that exhibit either obligate or facultative diapause, there are mothers that utilize both at the same time. Marsupial moms combine facultative and obligate diapause, and the most widely studied is the tammar wallaby. Like mice, tammar females experience ovulation directly after giving birth. They copulate and create a new embryo, which will enter facultative diapause.

The facultative signal for the newly minted embryo to go into suspended animation is the stimulus of the newborn in the mom's pouch. In other words, the suckling actions of the newborn wallaby sends a neural signal to the mother's hypothalamus that functions to prevent further development and birth of the second embryo. Wallabies give birth to their underdeveloped offspring in late January, the end of summer in the Southern Hemisphere. However, if something happens to the pouch newborn before May, for example, if the stimulus of the newborn suckling in the mother's pouch is removed, the diapaused embryo will reanimate, begin to develop, and be born approximately three weeks later. If the first offspring, the pouch young, remain healthy and developing or if something happens to it after the critical period in May, the mother will keep the encysted embryo in suspended animation until the following winter solstice (summer in Australia) the following December.

Once the solstice has occurred, the decreasing day-lengths are the obligate cue to reanimate the stored embryo. The embryo will be born approximately three weeks later, at which point the mother ovulates, copulates, and the cycle begins again. The tammar wallaby has among the longest phases of embryonic diapause of any mammalian species, at eleven months.

Bats are another mammalian group that experiences reproductive delays; however, in these species there isn't a diapausing embryo. Bat mothers can store sperm for long periods of time, something that is commonplace in the invertebrate world but much less common in mammals and primates. The sperm then have time to engage in their own cryptic battles within her storage organ. Mating in bats isn't always consensual. Males of many temperate species will often mate with females while they are suspended in a state of torpor (semi-hibernation), so females do not have a say in the matter at all. The last laugh however, comes in the form of something called *cryptic female choice* (CFC) in which aspects of a female's physiology allow for the postcoital selection of the most desirable sperm. Fertilization can occur many months after copulation, which provides ample time for sperm to duke it out in a female's storage chamber prior to the winner being allowed access to a female's precious egg cargo.

The unfortunate truth for many females in the animal kingdom, not just the bat mothers mentioned above, is that they must deal with manipulation, deception, and violence from their male counterparts. Sexual coercion is commonplace in all animal groups—females often have no choice about who is having sex with them. However, many mechanisms of CFC have evolved in direct response to this circumstance. Unfortunately, CFC, along with both facultative and obligate embryonic diapause, is difficult to observe and test. The notion of cryptic female choice is especially difficult to tease apart from sperm competition. This is an emerging field of study, and biologists are learning more about these systems all the time.

A compelling example of CFC occurs in common house mice. Females adjust the "fertilizability" of their ova depending on the social conditions in which they were reared. These moms are able to physiologically alter the external properties of their eggs to make them more or less fertilizable by sperms. Why would they want to do this? Ovary defensiveness is one mechanism by which females can limit polyspermy, or the incidence of more than one sperm entering an egg. Polyspermy is deadly to an embryo, so it's not surprising that females would have evolved a mechanism by which to avoid it. However, when females are raised in a polyandrous scenario—in which there are many males and lots of different sperms around—her eggs become more resistant to penetration by sperm. Conversely, when females are raised in a monogamous scenario with only one male, their eggs are less resistant to sperm penetration. Having a more resistant egg when there is more likely to be a high level of sperm competition in her reproductive tract allows for fertilization by only the most competent and competitive sperm—which is clearly advantageous for the female. Regardless of the number and identity of males that have mated with her, the female mouse maintains control over some aspects of her reproductive environment and allows them to work in her favor.

Maternal control over egg fertilization and the course of embryonic development means that the establishment of motherhood is a far more complicated task than simply sperm meeting egg. The evolutionary process has dealt mammalian females some powerful cards to play, although by their very nature these biological processes are hidden and difficult for the outside world to observe. As biologists develop greater tools by which to study it, we can expect to understand more about how mothers affect the process of embryonic development long before a pregnancy is established.

7

A CHILD IS BORN

So you've managed to find a mate, to copulate with that mate, and to gestate an offspring to term. Now what? The process of giving birth remains unpredictable in the human species. Although we like to pretend otherwise, we really don't have any kind of handle at all on how it's going to go, even for subsequent births in the same female. This has led many women to opt for voluntary cesarean sections. In fact, 32.2% of births in the USA in 2014 were by cesarean section, and the number will almost certainly only continue to climb.

It has been predicted that the incidence of cesarean section births in the Western world is sufficiently high to have an effect on the evolution our species. Although this sounds a little astonishing, it makes a lot of sense when you consider the specifics of the leading hypotheses surrounding the birth process in humans. Pelvic structure is a major limiting factor in the size our offspring attain before birth, and it's also one of the main reasons why the birth process in humans is so darn difficult.

According to the *obstetric dilemma hypothesis*, there is a limit to the possible dimensions of our birth canal due to our bipedal stance. In other words, because we walk upright, our birth canals are constrained to certain dimensions, and our pelvises are currently at the upper limit of what is possible. Any increase in the breadth of our birth canal would have a negative impact on our bipedal posture and will therefore not be evolutionarily selected. There is natural variation in both an individual woman's pelvic measurements and in the head and shoulder size of the individual baby within them. If a woman's pelvic measurements are too small to naturally birth a full-term baby or if that baby is naturally too large to fit through its own mother's birth canal, or both, it is likely that mother and/or baby would die during the birthing process without some kind of intervention. This is where the cesarean section comes in, making it possible for a woman with a tiny pelvis to birth a normal-sized child and making it possible for a normal-pelvised woman to birth an overly large infant. In this way, the genes that allow for this kind of maladaptive scenario remain alive and well in the human population, such that tiny-pelvised moms make more tiny-pelvised babies or massive babies grow up to have massive babies of their own. Genes that would otherwise have been selected against (i.e. both the mother and father dying during childbirth) now remain in the gene pool of our species and keep the "undesirable" characteristics alive and well. In fact, human moms are paving a road in the opposite direction of the evolutionary process when you consider that some women opt out of childbirth to avoid the pain associated with it. In addition to facilitating the maintenance of tiny pelvises in our populations, the use of cesarean sections could potentially allow for successful gestation of babies with bigger heads (brains). Imagine a human subspecies that has a superior cranial capacity but is unable to naturally give birth . . . it's not out of the realm of possibility.

A similar kind of argument could be made for all kinds of undesirable attributes in our species. We have developed any number of medical interventions that serve the ultimate purpose of keeping our gene pool riddled with evolutionarily undesirable characteristics—from poor eyesight to allergies, disease, and certain disabilities. Westernized medicine allows us to keep genes in our population that would otherwise be selected against.

It's not only difficult for a female to give birth alone, it's downright dangerous. Humans are usually described as a cooperatively breeding species because we do not undertake the birth process alone. There are several unique characteristics that make it necessary to have help. First, as mentioned, the human birth canal size is limited by the fact that we walk upright. The evolutionary cost of having a wider birth canal and easier birthing process is too high, much to the dismay of any woman who has ever given birth without pain medication.

There are three planes within the human female pelvis through which a baby must pass on its way into the world. First, the entryway into the pelvis—officially termed the *pelvic inlet*—is the widest part of the passage in a lateral orientation. The baby then enters the *midplane*, the narrowest part of its pelvic journey, before entering the *outlet*, which is widened in an anterior-posterior orientation and positioned perpendicularly to the pelvic inlet. What does this very complex and lengthy journey mean, other than the obvious excruciating pain for the mother? It means that the baby must twist around while navigating the birth canal, and it emerges into the outside world with its head facing down—with its head facing away from the mother. It is this fetal rotation that is a key factor in the extreme difficulty of childbirth in our species.

This strange orientation of our offspring during the birthing process is called "anterior occiput," and it has some pragmatic implications for the birthing rituals observed in our species. Since the mother cannot get an immediate look at the baby's face, the first thing to emerge if everything is proceeding as it should, she is unable to determine if the baby is breathing, and she is therefore unable to immediately remove any mucus from the baby's airways or adjust an umbilical cord that has become lodged around the baby's neck. In addition, the mother cannot assist in her own offspring's birth by reaching down and pulling the baby from the birth canal because the baby would be strained against its natural line of flexion and would likely become injured in the process.

In addition to the constraints of our pelvic dimensions, human babies are born with considerably large heads and wide shoulders. Called "encephalization," the heads of human babies are markedly larger at birth than those of any other primate. It's not terribly surprising that the evolution of massive brains is characteristic to the species; however, humans remain constrained, at least while natural childbirth is still on the table, in the size of brain that can squeeze through the birth canal. Human babies are born with only approximately 30% of their brain tissue, which means that we must achieve more brain growth postnatally than any other primate species. Chimpanzees are born with approximately 40% of their brain tissue; the final size of a chimp's brain in comparison with a human's indicates that human babies would need to gestate for an additional seven months to achieve the same proportion of their brain tissue at birth. The bottom line is that humans are evolutionarily constrained in a few intimately connected ways. Humans exist at the maximum possible dimensions for both pelvic size and brain size at birth, which makes the birthing process extremely tough—tougher than for most other mammals.

For all these reasons, human birth has always required a helper. In most societies, including traditional and tribal populations as well as westernized societies, that helper is usually not the father. Even for those lucky mothers whose dutiful partners were "catching" the babies as they emerged, the father is generally nothing more than an accessory in the delivery room. This perhaps explains why in most cases in the mammalian and primate worlds, fathers are long gone before the birthing process even starts. Once his sperm has successfully fertilized an egg, his job is essentially done. For most females in the animal kingdom, it is the assistance of another female that is the most helpful and supportive during childbirth. Most often in the primate world, they are also mothers that have given birth themselves.

Female-to-female interactions are a key factor in the birthing process in all primates, which makes a lot of sense for a variety of social and physiological reasons. Having support during such a vulnerable time in one's life is undoubtedly positive, and having the support of a female who is already been through the ordeal makes them qualified to provide more appropriate care. In humans, this kind of social support has been linked to shorter labor times and to improved maternal care if it persists after birth. In addition, human mothers report lower incidence of psychological anxiety if they are accompanied by a female supporter who has been there before.

The birthing process in great apes is rarely witnessed. In fact, there are fewer than ten accounts of such observations in the wild that are completely documented in the scientific literature. The scarcity of such observations has been attributed to many factors including that many primates primarily give birth at night, outside of watchful eyes of predators and potentially infanticidal males. Although birth in all primates is associated with pain, it's not the same level of discomfort felt by birthing females in our species, and most make only subdued vocalizations. Birth is characterized by a low level of movement and may often occur quickly without much sign of the impending event. All of these reasons make the process very difficult to observe. In recent years, however, a few accounts of the birthing process in our closest relatives—chimpanzees and bonobos—have been recorded in the wild. Where they have been witnessed, both bonobo and chimp mothers in labor have been observed to build day nests in a tree prior to giving birth within them. It doesn't seem terribly surprising that these females are literally nesting to prepare for the birth of their offspring.

Both the chimp and bonobo moms had females with them during their births. Attendants stayed close to the birthing female, grooming her or even holding and cleaning the baby after delivery. Many female apes help to deliver their own babies by reaching down and freeing them from the birth canal. The physical posturing of both mom and baby lends itself to this possibility more so than in humans. There are also cases in both Old World and New World monkeys to suggest that attending females can assist in bringing the baby out of the birth canal. There could be many reasons for another female to help, including in cases with first-time mothers or with high-risk births. If they are able, mothers of most mammals and primates immediately attend to their babies once they've been born. They are cleaned, comforted, and held. Suckling generally begins within minutes of birth. Mothers clean up or ingest the amniotic fluid, and most often ingest the placenta. In the recently documented births of both the bonobo and chimpanzee mentioned earlier, mothers not only ingested their placentas directly after birth, they broke off pieces to share with the attending females that provided support during the process.

Most human mothers frown upon eating the placenta—called placentophagy—but well-founded hypotheses account for its widespread practice in other primates, and most other mammals too. Placentophagy and the associated ingestion of the amniotic fluid and membranes are observed in most mammalian groups including rodents, lagomorphs, ungulates, carnivores, and primates. In fact, the only species where afterbirth ingestion is not widely observed are the aquatic cetaceans, whose amniotic fluid is immediately diluted in the ocean, and whose moms are immediately tasked with keeping their babies at the surface of the water to breathe and are therefore not able to immediately consume the placenta, and humans. Perhaps instead of asking why animals *do* consume the placenta, a better way to pose the question is to ask why humans *don't*, since there are so many direct and indirect benefits to be gained from doing so.

Consuming all products of the afterbirth may serve an anti-predation or anti-infanticide function because it gets rid of the evidence that an infant was born; however, this doesn't explain the widespread occurrence of slow and steady placentophagy in individuals that give birth in trees and may not be susceptible to predators, or to other mobile species who could simply walk away from the afterbirth instead of hiding the evidence. Another idea about afterbirth consumption is that the placenta is highly nutritious and could replace energy lost to the mother in the act of giving birth. This is most definitely a possibility, and sharing this important tissue is likely to reiterate a supportive and positive relationship between the birthing mother and her attendants. It's reported that a tendency toward "voracious carnivory" occurs in many post-birth females—even if they are normally herbivorous. Perhaps this has to do with the high energetic expenditure of birthing an infant, and that certainly makes sense. However, some kind of primal desire to eat flesh really speaks more to an evolutionarily defined process that is meant to restore specific hormones or bioactive molecules back to mom. Indeed, some ethologists have suggested that the consumption of the placenta plays a role in solidifying the connection between mother and newborn via increasing maternal brain opioid circuitry and suppressing postpartum depression.

Something called the *placental opioid enhancing factor* (POEF) facilitates the functioning of opiates in the mother's body that are naturally released during the birthing process. POEF is also present in the amniotic fluid—which many animal mothers consume prior to delivering the baby. Moms who have done this therefore have an increased level of functioning in their opiate receptors, which means that birth could be less painful overall. Some literal food for thought.

Much research suggests that important olfactory signals within the amniotic fluid function to stimulate maternal behaviors. Studies in several mammalian species (rats, mice, rabbits, and humans) suggest that the function of olfaction or chemical signaling between mother and baby is critically important to the neurogenesis of tissues relating directly to maternal care. Many mammalian moms immerse themselves in the amniotic fluid or use it to mark their kin after birth; it's an imperative part of the physiological and behavioral rituals at the commencement of motherhood.

The active labor and delivery process in the bulk of non-human primate species is quite short, ranging from just a few minutes to five hours. Females undoubtedly experience a high level of discomfort and pain associated with birth, as is evident in their restlessness and vocalizations; however, for many of the reasons discussed above, the long labor and delivery process in our own species is unique among primates. But we definitely don't have it the worst.

Think back to those female brown hyenas and their pseudopenises or penile-clitoris with the capabilities of erection and relaxation that make them externally indistinguishable from males. Female brown hyenas give birth through those penises, and the process is far from pretty. The pseudopenis is three-quarters to one inch in diameter—which stretches to 2 ¾ to 3 ¼ inches, the approximate head circumference of the pups, as the pups emerge down the canal—and just over 8½ inches in length; pups are generally between 2.2 and 3.3 pounds in weight. This is obviously a fairly large size differential, and a high level of pain is associated with process of labor in hyenas. Females experience almost constant contractions averaging an hour prior to birth; they change positions often and frequently assume a squatting position or attempt to prop themselves onto a tree or other solid structure for support. A specialized hormone called relaxin is prevalent in a female's tissues shortly before she is due to give birth. Relaxin helps to promote the elasticity of the clitoral tissue during this extremely difficult process.

Biologists report an astonishingly high level of physical damage to mothers and babies alike, including a high rate of death (up to 40%) for first-time moms and pups. The primary cause of death for pups is usually suffocation, as the time it takes to travel down such a long and thin tube can simply be too long to sustain life. There is also significant and potentially fatal tearing for mothers. However, once a mother has given birth once, it gets much easier, because the glans (or head) of the penile clitoris invariably rips open during the birth of a female's first pups, alive or not, and this facilitates subsequent births.

Such difficult births are far from normal in most other mammalian species, although human births do run a close second to the horrors of hyena parturition. For the bulk of the animal kingdom, the process of birthing offspring is not as difficult or traumatic as it is for humans. Did the difficulty of parturition in our species contribute to the evolution of smaller families? What about the evolution of grandmothering? It's possible that our species was naturally predisposed to cease our reproductive actions at a relatively early age simply because it's so difficult. Rest assured, for most other mammals, and animals period, birth is undoubtedly uncomfortable but far from what we encounter.

8

NEWBORNS AND NIGHTMARES

Physiological, social, and environmental elements have influenced the evolution of the immense diversity in maternal behavior across the mammalian order. There are two general phases of the establishment of maternal care in primates—the initial bonding or recognition phase and the maintenance or persistent attraction phase. The first of these is essentially what occurs in the moments directly following birth—in which mom and baby get their introduction to one another. This is a profoundly important period for establishing the foundation of the mother-infant bond, and it has both physiological and emotional components. The maintenance or maternal memory phase occurs in the weeks and months following birth, after hormones have reached critical levels within mom's body. No further hormonal input is required to maintain maternal behavior; it is sufficiently ingrained in mom's physiology and mind. In other words, a mother has transformed into a different animal than she was before (if this was her first offspring); she is a mother now, and irreversible changes have taken place in her body and brain.

The recognition phase can be further subdivided into either selective or non-selective. Non-selective moms typically give birth to underdeveloped young, and they provide maternal care to any newborn in their vicinity. Non-selective care is directed toward a general offspring cue rather than a specific one. Rodents like rats and mice exhibit non-selectivity toward newborns (remember the paternally expressed imprinted genes?), and females combine their litters and provide care equally to all pups. As long as all pups in a litter or nursery (combined litters of several different mothers) have their needs met, everybody wins. A rodent mother does not have a lot to lose by providing maternal care to all pups in her immediate vicinity, and if she happens to be caring for a few that aren't biologically hers, chances are another mom in the area is doing the same thing, so it all evens out.

On the other end of the spectrum is selective recognition, in which the mother immediately learns the olfactory signature of her own offspring and subsequently directs all her care exclusively toward it. Selective recognition generally occurs in either of two situations—first, when mothers birth offspring that are precocial (who are in a more advanced state of development at birth, like ungulates and sheep), and second, when offspring are semi-mobile, as with most primates. Precocial newborns have a full suite of sensory capabilities at birth; they emerge from the womb, and generally get up and walk or swim immediately. Mothers almost immediately recognize their own offspring and provide their maternal care exclusively. In the case of selective recognition, the costs of providing maternal care are generally higher, and mothers have only one or two offspring instead of a litter. These factors make it important for mothers to have some means of distinguishing their own offspring from all others.

Pre-weaning vulnerability to predation pressure is a major factor for all mammalian species (except—arguably—humans), and mothers have their work cut out for them to keep their defenseless newborns from meeting their deaths at the hands of predators. Mammals who give birth out in the open, either on land or in the sea, are particularly vulnerable to visual predators. These kinds of animals, like even-toed ungulates (Artiodactyls, including giraffes, deer, hippos, and whales) and odd-toed ungulates (Perissodactyls, including zebras and rhinos) generally give birth to a small number of relatively large offspring at infrequent intervals. Since these newborns are precocial, they are able to follow mom directly after birth.

Many ungulate species are monocotous, meaning that they have only one calf at a time. Herbivorous grazers like deer, giraffe, and sheep are under constant predation pressure from large carnivores. Moms of these species generally take on one of two different strategies when it comes to protecting their newborns in their first few weeks of life: hiding or following. In "following" species such as reindeer, newborns immediately follow their mother after birth and remain in very close contact with her. Newborns are almost immediately mobile, and they can rely on their mothers for protection just as long as they can keep up. Large groups of individuals may remain together to source out food, and they generally cover significant amounts of area to find enough. This means that any newborns must keep pace with the group or they will be left behind to die. In these following species, moms are always on the move for both resource gathering and predator avoidance, and both mothers and newborns are equipped with mechanisms to recognize each other. A combination of olfactory and vocal cues allows mom and infant to remain in close contact with each other at all times.

On the other hand, "hiding" species tuck away their newborns in patches of vegetation to keep them safe from predation. For example, directly after birth, fallow deer moms will find a safe patch of vegetation where her neonate will hide—and remain silent—for the next two or three weeks. Mothers spend most of their time away from the place where their offspring are hiding, though they come back at intermittent intervals to feed them. However, the females don't always recognize the exact spots where their offspring are hiding. Mothers approach the approximate location of their fawns and begin to vocalize. Each mother has a very distinctive vocal signature, which is recognized by the pups. Mom makes the right sound, and baby comes out from its hiding place to feed when it hears its own mother's call. Mothers are unable to recognize their own fawns based on call signature alone. Unlike in the "following" ungulate species where both mom and newborn can recognize each other, in "hiding" species, it's the ability of the pup to discern the unique call of his or her mother that results in reunification. It may seem counterintuitive, but it's evolutionarily more important for the fawn to be able to recognize its mom than it is for the mom to recognize the fawn. The mom could lose one offspring as a result of lack of recognition, but the fawn could lose its entire existence. After the initial period of hiding, fawns join their mothers and larger groups of females and other fawns of both sexes. Young deer are rarely in the company of sexually mature males.

Although they employ neither a "hiding" nor a "following" strategy, bottlenose dolphin moms give birth in an open environment as well. Newborn dolphins remain very close to their mothers for the first few weeks after birth, similar to the "following" strategy of ungulates. Something called *echelon swimming* takes place for at least the first week, in which the newborn receives physical support from its mother. The neonate is led along by its mother, almost draped over her head area. The two come to the surface to breathe in synch with one another, and this extra support helps the calf develop its own coordination of moving, breathing, and diving. Dolphin newborns represent a unique category of animal baby because they are air-breathing, but are born in, and live entirely in, water. In addition to the shock of getting born, dolphins also must learn to swim and learn when and when not to breathe. This makes them more precocial than terrestrial counterparts, because without these immediate skills they would certainly drown. However, dolphins share many characteristics with primates—things like a long lifespan and slow life history. So, although baby dolphins are born with the innate ability to survive, much like human babies they still have a lot of cognitive developing to do after birth.

After a week of echelon swimming, newborn dolphins assume a swimming position directly underneath the mother's abdomen. This is called the "infant position," and newborns remain here for a few more weeks as they gain increasing independence from their mom. The early transition from a position beside mom's head to the spot on her abdomen is reflective of the developing abilities of the baby. When its swimming becomes stronger, it no longer needs the extra support. The newborn position still affords the baby protection and proximity to mom's mammary slits, but it doesn't need to be propped up any more.

Unsurprisingly, the most vulnerable stage of life for most whale and dolphin infants is directly after birth. This period is marked by repeated bouts of fast swimming in dolphins, whales, and many others, which could be the reason behind the evolution of the echelon position. Moms help their neonates learn to swim, remain buoyant, and learn to breathe, all while moving at a fast pace to avoid predation.

In addition to remaining very close to their moms, newborns rub against them and pet them, most often near their heads. Many dolphin moms experience attempted sexual advances from their infant sons within three weeks of their birth. Bottle nosed dolphins are hypersexual creatures that engage in sexual activities several times per hour as adults, so this observation is not particularly surprising, nor is the fact that erections in newborn male bottlenose dolphins are often observed within two days of birth.

After the first month, dolphin babies begin to socialize and learn to forage away from their mothers. The amount of time they spend away from them is related to age, and moms encourage independence of their offspring as they often must make foraging trips of substantial distance. Several mothers will group their offspring together to create larger groups of moms and babies, which serves many purposes. First, it keeps the juveniles safe from predation and from unwanted sexual advances of adult males (groups are comprised of females and juveniles only), and second, it allows the juveniles to socially engage with each other and to learn to play, forage, and communicate.

Unlike animal moms that give birth out in the open, there are many other species who give birth in much more protected spaces. Many make burrows, hiding places, or nests in which they bring their offspring into the world, and this provides an immediate shelter from predation and the elements. Nesting moms generally give birth to larger litters of offspring that are less developed. Animals such as carnivores (wolves, tigers, weasels, raccoons), insectivores (shrews, moles, hedgehogs), and lagomorphs (hares and rabbits) have a large number of small infants. Although we often joke about rabbits having more sexual relations than most other creatures, they actually don't. However, their physiology is such that in adult female rabbits the process of ovulation is linked to the process of copulation. Each time a female rabbit copulates, she gets pregnant. In the wild, rabbit litters contain three to six kits; however, as with most mammals, maternal behavior begins far before parturition.

The gestational period for a rabbit is around thirty days. Approximately one week before giving birth, a pregnant female will begin to dig an underground chamber for her kits. Within the chamber, she constructs a soft nest of vegetation, loose fur plucked from her own coat, and fecal pellets. During the parturition process, the mother rabbit gives birth to a new kit approximately once per minute for the three to six kits total. She will stand over her nest and briefly lick the newborns as they emerge. She also consumes each placenta rapidly, despite being otherwise exclusively herbivorous. Directly after the birth of her last kit, the mother leaves the nest and seals up the burrow from the outside world. For the next two weeks she will only visit her newborns *once* per day, for the exclusive act of feeding. Rabbit moms do *not* cuddle, groom, or clean their offspring—and if one of them happens to stray from the burrow, it will not be retrieved. Each burrow-bound nursing bout lasts for only approximately three to four minutes, and this is the only contact that the mom has with her babies each day. It's an intriguing form of absentee mothering that is exclusive to the lagomorph group. Rabbit moms have essentially decoupled maternal care from maternal presence—which is shocking news to anthropomorphic humans that would love for mommy rabbits to be as cuddly as they are cute. (Take *that*, Easter Bunny!)

Although it seems odd that a new mother would provide so little care to her offspring, it's likely to have evolved due to the extreme vulnerability of rabbits to predation pressure. Each time a mom enters or exits the burrow, she risks disclosing the location of her offspring, and herself, to nearby predators. The extremely low level of infant visitation in wild rabbits has resulted from the need for mothers to keep the location of their newborns a secret.

Indeed, rabbit moms are alert to their infants' distress calls and to the potential of predatory threats.

An indirect effect of this absentee mom strategy is that the bunny newborns form very close relationships to one another in their first few weeks of life. Instead of relying on their mother for warmth, thermoregulation, or company, neonate rabbits rely on each other. Sibling interactions within the burrow form most of an infant's social development, although competition between them for mom's milk is quite severe. Approximately 20% of wild rabbit neonates die of starvation within the first postpartum week of life. Infant body mass is a good predictor of their ability to compete for milk; larger offspring obtain more milk because they can out-compete smaller siblings. Their position within mom's uterus prior to birth can affect the ultimate body size of the infants. Those at the ends of the uterine horns (as opposed to being in the center of the uterus) tend to have a larger body mass at birth.

It really seems like the deck is stacked against neonate rabbits. However, although they are fed infrequently, housed in a dark underground burrow, and blind for the first nine to ten days, the newborns are still able to obtain enough nutrition. When mom enters the burrow, they immediately find her nipples through distinct pheromonal (actually called the "nipple search pheromone") and olfactory cues, and they jostle with each other for a good position to feed. The newborns have an internal circadian rhythm that complements the maternal signals by creating an anticipation for the mother just prior to her arrival for feeding. During or directly after nursing, pups urinate and then dry themselves off while snuggling back down into the soft material of the nest. It's as though mom is little more than a food vending machine; her tiny pups get their share and immediately look back to each other for comfort.

Aspects of the mother's diet are discernible from her milk, and so her newborns develop a sense of what foods to begin eating once they make the transition to solids. The transition to solid foods begins while the kits are still within the burrow. At around day twelve, the young rabbits begin to nibble at their mom's fecal pellets—which functions to build up their gut microbiomes, in much the same way as baby koalas ingest their mother's feces soon after birth. They then begin to consume the nest material, mostly dried vegetation, until they fully wean and leave the burrow at approximately three weeks after birth. Then mothers abruptly stop nursing their kits, even going so far as to attack kits that try and suckle after this time. Most often, rabbit mothers are already pregnant again, so they will be focused on creating a new nest for the next crop of neonates to come along.

There is also a third general category comprised of moms that give birth to semi-precocial offspring (in a quiet setting but not specifically out in the open or in a natal burrow) that require an extremely high level of care. For example, primate neonates are most often transported with the mother from birth. Very soon after they make their first appearance in the world, newborn primates are able to cling to their mothers, and they rely on them for *all* aspects of their existence for a lengthy period of time. Primate mothers vary greatly in terms of the kind of mothering they provide to their first infant versus subsequent ones. Unlike many other mammalian moms that exhibit a full suite of maternal behavior from the get go, primate moms have a *lot* of learning to do with their firstborn. For many aspects of newborn care, it's a matter of trial and error.

The process of motherhood is set in motion by the complicated series of physiological events that begin as part of the maternal transformation. The mother's endocrine system (which is the body's system of signaling via hormones and neurotransmitters) undergoes a substantial set of changes throughout pregnancy, parturition, and the peripartum (just after birth) periods which facilitates the biological conversion from nulliparity (no offspring) to primiparity (first offspring). In addition to the familiar hormonal signals—both gonadal hormones like estrogen and progesterone, and adrenal hormones like corticosterone—there are monoamine neurotransmitters and opiates that have distinct roles in the metamorphosis of motherhood. The field of behavioral endocrinology aims not only to pinpoint the specific physiological pathways and acting mechanisms of each chemical signal, but also to trace the resulting behaviors that occur. In other words: how do specific hormonal and chemical signals impact how a female behaves with respect to motherhood?

There are some general trends when it comes to hormonal signaling; however, there are also substantial caveats to their interpretation. One of the most difficult things about trying to pattern maternal behaviors from certain physiological changes is that there are many outside factors that cannot be controlled for. There are countless circumstances for each mom—both ecological and physiological—that make it difficult to fit maternal behavior into neat and tidy categories. Mammalian and primate maternal physiology is endlessly complex, and it's worth keeping that in mind when observing general trends.

For most female primates, the most well defined maternal-functioning hormones are estrogens and progesterones. These are the gonadal hormones, and they are among the most conspicuous in matters of reproduction and motherhood. These hormones are significantly elevated during pregnancy, especially during the latter half. They significantly decrease at parturition and remain in low concentration and acyclic during the postpartum phase of lactation and lactational anovulation (the lack of female ovulatory cycling that occurs during lactation). Estrogen and progesterone play substantial roles in the onset of maternal behavior in both primate and non-primate mammals. Females tend to show increasing interest in babies and infants throughout pregnancy as a direct result of increased gonadal hormones. For example, female pigtail macaques and many other female primates show marked interest in the newly birthed offspring of other females while they are pregnant. (It's interesting to note that these same processes are alive and well in the human primate—it's part physiology, part curiosity, and part excitement.) As circulating levels of estrogens peak during the last weeks before parturition, female macaques and baboons show significantly higher levels of infant handling; however, in marmosets the trend appears to be just the opposite. A negative correlation exists between estrogen levels and the onset of maternal behavior in these New World monkeys.

Prolactin is another hormone that is critically important for the onset of maternal physiology, as it is the primary signal for the stimulation of lactation. However, not much is known about the potential effects of prolactin on maternal behavior. It's suspected that prolactin may be an important hormone for its role in the onset of allomaternal behavior in several primates including marmosets, tamarins, titi monkeys, and squirrel monkeys.

Oxytocin (OT) is a hormone that is released as an automatic response to the vaginocervical stimulation that occurs during birth. The specific stimulation in the vagina and cervix is what instigates the initial release of oxytocin and associated maternal behaviors. Subsequent doses of OT are released as the uterus contracts to its original size and as nipples are stimulated during suckling. The role of OT in the bonding process between mother and baby is massive, as it is during other pair-bonding processes like falling in love. In all mammals, OT acts to promote social bonding, to modulate sexual behavior, and to reduce stress responsiveness and anxiety. Although there is considerable variability in the level and outcome of OT injection, its overall importance in bond formation is key. Not just generic bond formation either: the bonds between lovers or between mother and child, the most important and affectionate bonds, involve the secretion of OT from the brain.

Other chemicals, including opioids and monoamine neurotrans-mitters like serotonin, also influence the early stages of maternal behavior. Opioids influence affiliative and bonding behaviors in mammals and primates. In fact, the *brain opioid theory* of social attachment suggests that the role of opiates like beta endorphin in the mother-offspring attachment is substantial. Circulating levels of beta endorphin released from both the brain and placenta increase during pregnancy to a maximum level just prior to birth in both rats and humans. It is therefore suspected to have a role in preparing the mother for developing a strong, affectionate attach-ment to her offspring across the mammalian class. Serotonin plays a major role in preparing females for motherhood. The level of serotonin in the brain affects anxiety and impulsivity levels in new mothers, which has direct influence on the offspring. Conversely, maternal rejection of an infant can affect development in the brain of the infant. The long-term effects of a poorly developed serotonin system are still under investigation (as are the relationships of many neurotransmitters to maternal behavior); however, it is likely that it could lead to a repeated cycle of infant abandonment in those that were themselves abandoned.

The number of hormones, neuropeptides, and other endocrine factors and their complex relationships to maternal instinct and behavior make this a difficult field of study. Broad-reaching, clear conclusions are difficult to come by. However, the bottom line is that there are abrupt physical changes that mark the transition of females from being offspring-aversive virgins to offspring-seeking nurturers. Importantly, there are also cues in newborn offspring themselves that serve to stimulate adult females to nurture and care for them. In the best-case scenario, the cues from baby serve to complement the cues from mom to create an optimal level of care for the former.

The fact that some cues come directly from the baby means that it's possible to induce maternal behaviors in virgin females by repeatedly exposing them to newborn pups. This means that not all maternal behaviors are physiologically or hormonally based, which is an important fact because it shows that *all* females are predisposed toward maternal tendencies when the correct stimuli are received. This is evident in behavior of several mammals including rats, who begin to exhibit maternal behavior after approximately seven days (termed the sensitization latency) when experimentally exposed to newborn pups. After sufficient time, a virgin female can be induced to retrieve, protect, and even assume a lactation posture over pups that aren't her own, even though she cannot provide them with milk. The maternal scaffolding is in place to the extent that certain behaviors may be induced in this way.

The last of the endocrine changes that should be mentioned are the changes to the hormones of the hypothalamic-pituitary-adrenal (HPA) axis, which plays key roles in metabolic processes as well as regulating physiological and behavioral responses to stress. The specific neuroendocrine mechanisms by which new mothers respond to stress remain elusive, but they most certainly involve the hormones of the HPA axis like cortisol, corticosterone, glucocorticoids (GC), and corticotrophin-releasing-hormone (CRH). In primates such as gorillas, bonobos, chimpanzees, and humans, basal concentrations of circulating stress hormones rise throughout the course of pregnancy, increasingly during the latter half. Circulating cortisol increases the arousal and responsiveness of moms-to-be, which generally results in them being more careful and less adventurous during late pregnancy. At birth, there is a spike of GC, which is thought to be crucial for the timing of the birth itself.

In some primate species (such as Japanese macaques), GCs circulating in mom's pregnant body mean that toward the end of pregnancy, an expectant mother's response to stress becomes severely attenuated. It's essentially the mother's own physiology working to dampen the reactivity of the part of the brain associated with stress. Why would such a process be advantageous to an expectant mom? Stress hormones can cross the placental/fetal interface and potentially harm the developing baby, so it makes sense for the mother's body to go into "low stress mode" during the last part of pregnancy so that this can be avoided. That's not to say that mom isn't on her guard due to the effects of slowly increasing cortisol levels in her body during pregnancy. It's a complementary process which sees mom being simultaneously more careful and alert and yet less stressed out while she is in the most vulnerable stages of pregnancy. These are termed "adaptive emotional processes" because they enhance and sustain a mother's motivation to invest in and protect her infants. This is a reinforcement of the evolutionary tendency toward ensuring that mothers stay safe, quiet, and most likely hidden or inconspicuous during their last phase of pregnancy. This state of high vigilance and anxiety peaks just prior to birth and then markedly decreases during the lactational phase.

It's difficult to study the impact of stress hormones on pregnant, birthing, and new moms, and on their offspring in the wild; it's impossible to control for every mitigating factor that could lead to differences between two females of the same species. What scientists *do* know is that there are ways in which a pregnant female's body acts as a buffering temple for her offspring; however, that's not to say that things cannot go wrong . . . because they certainly do. Despite the natural buffers that are in place to protect developing infants from stress hormones, many pregnant primates experience stressful circumstances that are both beyond their ability to control and beyond their own bodies' shock-absorbing capacity. What then? When fetuses are directly exposed to stress hormones in utero, or babies are subjected to stress hormones through lactation, the scene is set for negative outcomes—both short and long term. The peripartum period represents a time of such massive change for females, both physiologically and emotionally, that renders them naturally vulnerable in a number of ways. When new primate mothers experience elevated stress, they are more likely to pass on its effects to their offspring. Stress-induced disregulation of emotions can easily interfere with maternal motivation and consequently maternal behavior. In fact, stress is a significant risk factor for postpartum depression in primates, from macaques to humans. More on that in a later chapter.

Even without a great deal of stress of complication to amp up the drama, the physiological and emotional happenings of a "normal" parturition is fascinating. Once a female mammal transitions from being nulliparous to being primiparous modifications to the brain and body occur that are irreversible and complete, and these changes—in combination with repeated stimulus of the baby and the initial maternal experience of caring for her own offspring—mean that mom is forever changed. There is a suite of physiological changes that occur in the maternal brain that facilitate the dramatic transition to motherhood. It's a little surreal to think that your own body is somehow conspiring against your own primary interests to make you less selfish and more ready to provide all your energy and strength toward the survival of another creature, but that's exactly what is happening. The immense plasticity exhibited in the mammalian maternal brain and body is among the most remarkable embodiments of the evolutionary process.

In addition to the neuroendocrine signaling processes outlined above, there are also physical changes that happen directly to the mother's brain. Such neurobiological changes begin during early pregnancy, and they continue gradually during gestation. At birth, the rash of additional physiological changes, like those detailed above, cue the onset of maternal behaviors that work in concert with the modifications that occurred throughout the pregnancy. Any mom will recognize the term "baby brain"—the forgetfulness or absent-mindedness of pregnant or early postpartum females. Baby brain is in fact a bona fide issue for all mammalian females. In human moms, brain volume during pregnancy decreases by approximately 4.3%, and it doesn't return to normal capacity until approximately six months after birth. Energies otherwise dedicated to the needs of the mother alone become redirected toward the offspring instead. Substantial changes occur in several brain areas including the hippocampus, amygdala, and the olfactory bulb. The immense collection of changes is called the *maternal circuitry* and this is assumed to be largely conserved across mammals.

In primates, the first phase after birth is generally characterized by an intense bonding period between mother and baby. Most primate species are characterized by long lifespans, slow life histories, and long periods of immature development. Such characteristics mean that mom is a very important leading force in her offsprings' lives for a long time. The overall pattern of maternal care in a specific group is reflective of the social system at play. For example, prosimian monkeys are small-bodied, nocturnal primates that are largely solitary. After birth, mothers will "park" their infants in a hiding spot or nest while they go off to forage. During her brief visits to her offspring, mom will feed and groom them at the nest. In New World monkeys like marmosets and tamarins, females usually give birth to twins. Extra help in terms of carrying her offspring is required right from the get go—and such help usually comes from dad or from older offspring. Infants are carried for weeks, months, or even years after birth, depending on the species. Many New World primates exhibit a cooperative breeding system, where not all females are reproductively active (see *Cooperative Creeds and Communal Cliques*).

Old World monkeys and apes have singleton births (one baby at a time), followed by the aforementioned intense period of maternal care and bonding. Lactation in primates can be a lengthy process—in most cases females provide milk to their babies for at least a year and sometimes several years (see *Mammal Moms and Milk*). Most primate species live in one-male groups with a polygynous mating system (macaques, baboons, monkeys) or live in large multi-male/multi-female groups where both males and females mate promiscuously and groupings of individuals are fluid in nature (bonobos and chimps). Males do not play a direct role in infant care; they instead protect the group from outside dangers.

Most of the physiological and emotional changes associated with pregnancy and parturition in primates result in interactive and motivated mothers who provide most of the care to their own newborns. They maintain contact with their infants for long periods of time and usually nurse on demand. During the first few postpartum weeks, mothers are in a constant state of nursing their offspring. Mothers also provide extensive grooming to newborns, constantly reinforcing their touch, presence, and bond with them. A key process by which new mothers make meaningful contact with their babies is through something called a *maternal gaze*. The formation of secure attachment bonds depends on sensitive maternal behavior, especially during the first few months of life. The maternal gaze has the power to convey a good deal of emotion, security, and love to a newborn during a critical phase of development. Indeed, internal oxytocin levels are associated with frequent infant gazing and quality of maternal care.

Infant apes are born with clinging reflexes intact, so it's rare to see a postpartum female without her baby clinging to her. As with humans, the first postpartum phase in great apes is a highly demanding job, twenty-four hours a day and seven days a week. This is an interesting point at which to examine our own species—one in which husbands and fathers are largely left out of the newborn process. Many dads play the alpha role, working a job outside of the home to provision his partner and offspring; generally, however, human dads aren't required as a first line of defense against predation or environmental stochasticity. Moms are left in the home with newborns, often feeling overwhelmed and exhausted, and dads head back to their jobs soon after their infant's birth, usually with a feeling of being left out of the process. Biologically speaking there isn't much a human dad can do; however, our species has come up with a myriad of ways in which we can manipulate the system and allow dad to bond with his newborn as well. Bottle feeding, cuddling, and providing primary care are just a few examples.

The early phases of motherhood are characterized by a high level of individuality in both maternal personality and parenting style. It's difficult if not impossible to classify all mothers as having the same measurable behaviors and effects on their offspring. Imagine trying to attempt such a consensus among human mothers! There are as many unique styles as there are mothers who are engaging in them—and, for animals with large brains, complex emotions, and social structures, it's no small feat to try to generalize about how new moms act in a particular context. In addition to the clear individualities in emotion and reactivity, there are also more specific characteristics to consider for a particular mother. Things like maternal age, parity, and social rank can all play a role, as well as environmental characteristics like resource availability, group size, and predation risk.

One way to obtain more detailed information on the process of motherhood in primates is to examine a particular population for a long period of time. There are semi-managed research populations that serve this exact purpose—to allow for the kind of long-term observation that would shed some light on more general aspects of motherhood and associated social behaviors. While these study populations inevitably involve some level of human contact, they remain a valuable resource for many kinds of biological data on primates like baboons, macaques, and vervet monkeys. For example, a population of *Rhesus macaques* on Cayo Santiago, a small island near Puerto Rico, was established in 1935 when 409 individuals were imported from India. Today, the population is well over 1,000 and well-studied by researchers from several scientific fields.

Collectively, ecological work on both semi-managed and completely natural primate populations has resulted in a few generalizations based on aspects of maternal protectiveness, warmth, and rejection. A mother's *protectiveness* is defined as the extent to which a mother maintains close proximity and contact with her infant, as well as the extent to which the infant is physically restrained. *Warmth* is defined as the extent to which mothers engage with their infants in cuddles, grooming, or nursing. *Rejection* is the extent to which mothers limit contact with their own offspring. Individual moms, regardless of species, tend to adopt a similar parenting style with consecutive offspring, further reiterating the individual nature of the maternal process.

Extreme cases of rejection can be considered abandonment, which is observed infrequently in primate moms. When it is observed, it's most likely to be perpetrated by a first-time mom who simply may not be aware of the severity of her actions. There are times when a low level of maternal rejection is a positive thing. For example, many chimpanzee moms are in almost constant contact with their offspring for the first several years of their lives. Some individuals remain in close contact with their mothers until well past weaning, even into adolescence. For these and other primates in which there is a long period of intense maternal investment, moms often initiate the weaning process through a rejection of their own infants. It's stressful for infants to experience this kind of dismissal from their own moms; however, it's necessary for them to begin to develop their own skills of exploration. Assuredly, many human moms can relate to this point. Primate mothers who are intensely protective of their infants tend to promote a sense of fear and trepidation in them, while mothers who are intensely rejecting have highly adventurous and curious infants that do not develop normal emotional attachments or fear. The latter tend to develop independence from their mothers at an earlier age and spend much more time associating with other conspecifics (members of the same species) and peers, while the former can be delayed in their acquisition of independence and remain cautious throughout their lives.

As with most things in motherhood (and life), some kind of middle ground is likely the most appropriate approach when it comes to the contrasting costs and benefits of being protective versus being rejecting.

9

DEPRESSION AND ABUSE:
NOT JUST A HUMAN PREDICAMENT

he peripartum period is formally defined as the last few weeks of pregnancy and the first few months after birth. It's generally a time of calm and quiet, as mothers are focused on direct care of their newborns and establishing the relationship. The high levels of anxiety experienced during the later phase of pregnancy give way to a low-key phase of direct maternal care, where neurobiological changes continue to occur and function toward the long-term aspects of maternal behavior and identity (the maintenance or long-term attraction phase). Under healthy conditions, the maternal brain undergoes a transformation that will be sustained through senescence and even through the entirety of her life. However, the maintenance phase also represents a time when mothers are particularly vulnerable to outside sources of stress. In a perfect world, mothers and their newborns would have long phases of quiet, uninterrupted time to recover from the birth, rest, and bond with one another. Such a utopia can be rare for some animals, including westernized humans, where birth may take place amid a myriad of extenuating factors like predation risk, environmental chaos, social rank, or localized illness.

In addition to the myriad of factors mentioned above, there are aspects of maternal care that are dependent on each individual mother's unique personality and on the developing personalities of their infants. Mothers need to learn about each individual infant's temperaments and specific needs—how they like to be held, cuddled, and fed—and how the moms like to do these things too. This type of exploration is most difficult for first-time moms, who may not immediately be comfortable holding or suckling an infant. For most primates, having one's first baby comes with a huge learning curve. It doesn't automatically feel natural to hold the baby or deal with its cries or whines; it takes some practice and getting used to. Perhaps not surprisingly, the incidence of infant abandonment is highest for first-time moms across many primate species. If conditions are suboptimal and a new mother is very young, it may even be biologically advantageous to "cut her losses" once she has given birth. Since newborn primates require such a high level of immediate care after birth, infant abandonment is a certain death sentence. This isn't to say that first-time moms are abandoning their offspring because they are depressed, it's most often a case of simply not knowing any better.

Multiparous primate moms, humans included, have a much higher success rate and a very low incidence of infant abandonment. Our long lifespans mean that we have an extended period to get this whole motherhood thing figured out to the best of our ability, even if we didn't get it all right the first time. Keep in mind that the suite of behaviors that moms learn, and that allow them to develop their own maternal style, is in *addition* to the physiological and neurological changes that have already occurred. It's the latter changes that allow for the establishment of the former behaviors. Essentially there are two main mothering forces at work here: physiological-induced mothering and experience-induced mothering. These are not exclusive of one another, and each can induce feedback to the other as well—which makes this a difficult phenomenon to study. Unlike in primates, rodent moms exhibit the full spectrum of maternal behaviors right from the get go (lucky them). First-time rodent moms are as competent as moms with previous litters, leading to the conclusion that maternal behaviors and responsiveness are under strict genetic control. This makes evolutionary sense seeing as most rodents have shorter lifespans and less opportunity to provide direct maternal care.

Outside factors can play a major role in the development of primate maternal behaviors as well. Things like the mother's age, number of previous offspring, and rank (whether they are dominant or subordinate) may also affect the level of care she is able to provide, and the kind of care that fits within her social identity. Environmental characteristics have the potential to completely change the ways in which mothers provide care to infants. As with most aspects of life in the animal kingdom, when resources are plentiful and easy to obtain, more energy and effort is available for activities other than foraging. However, if resources are difficult to obtain because of unforeseen circumstances, such as severe weather patterns or the threat of predators, mom has to make choices about how best to serve her young—and herself. In addition, mothers likely must incorporate aspects of social group structure into their everyday parenting decisions to maximize cohesion and avoid competition, infanticide, or kidnapping. So, while there are basic physical features of motherhood that are universal, there is a myriad of ways in which the kind of care that each individual mother provides is unique.

The physiological and neurological changes that occur during gestation are all meant to maximize survival of both mom and off-spring, but they also help to provide mom with a buffer against the profound hormonal changes that take place in her own body at par-turition. There is a rapid drop of estrogen and progesterone with the expulsion of the placenta, and corticosterone levels remain steady and somewhat elevated (as they naturally were during pregnancy). Corticosterone levels naturally begin to drop at the mid-lactation point. All these hormonal changes are essential for the onset of maternal behaviors, but for certain vulnerable individuals, they can induce symptoms of depression. It's somewhat difficult to clearly discuss what kinds of specific changes happen and the particular behaviors associated with them because there is the added effect of individual variation in behavioral, physiological, and environmental characteristics. Mammalian and primate moms are complicated creatures, and the care of their own offspring isn't simply a job that they are physically programmed to carry out (as may be the case with the rodent moms described earlier). New moms (whether cow, dolphin, baboon, or human) are vulnerable to the effects of stress, and they are at high risk of developing stress-related psychiatric disorders. One of the most common is postpartum depression (PPD), which is experienced by approximately 15 to 20% of women in westernized societies. Stress is a major risk factor for PPD, as it directly interferes with the physiological mechanisms in place to provide calm, uninterrupted care to newborns. Chronic high levels of corticosterone in any mammalian mom during pregnancy and early peripartum is associated with a reduction in provision of maternal care. Despite the vast levels of diversity that we observe in early motherhood among individuals, one thing is clear: stress and consequent physiological elevations of corticosterone in expectant and new moms interferes with the neurobiological processes that create normal maternal behavior.

As mentioned above, there are stress-induced changes to neurological structures and neurogenesis that interfere with the transition to motherhood, and this sets the scene for psychiatric disorders in new moms. Disruptions to the functionality of oxytocin (OT) are generally concurrent with increased stress levels in new moms too. Mothers that face higher stress levels are more likely to have lower levels of OT, which is correlated with less attentive parenting and even negative care. In other words, the physiological changes associated with postpartum depression are very real, and they reinforce each other in a cyclical way. Stress causes high corticosterone, which results in lower levels of oxytocin, which makes mom feel worse and more stressed, and so the cycle continues. Postpartum depression is not something to be dismissed or belittled.

Our own societal standards often make new moms feel inadequate and unable to ask for help. Most of our early check-ups and check-ins with medical professionals are completely focused on physiology and not at all on psychiatry. This is a difficult conundrum for human mothers, who are tasked with doing the "most natural" thing in the world in the most unnatural set of circumstances. The fact of the matter is, there are at least as many changes taking place in a new mom's brain as there are in the rest of her body, and so her mental health should be as highly prioritized as her physical health.

There are three categories of depressive behaviors that are potentially experienced by new moms. The first is the "baby blues," which generally occurs around the fifth day after giving birth. This is a response to the rapid fall in progesterone (which can be steeper for some moms than for others), and those who experience the baby blues have a greater chance of developing symptoms in the second category, which is an exacerbation of the baby blues into postpartum depression. PPD is defined as a depression which extends past the point of days or even weeks after giving birth. The incidence of developing PPD is higher for new moms who have experienced the baby blues, and it occurs in around 25% of women who have experienced severe symptoms of the latter. Due to the massive changes occurring in a mom's body, it's no wonder that moms are over twenty times more likely to develop the third category of depression—postpartum psychosis—during the first month after giving birth than they are at any other time of their lives. Our culture expects new moms to be thrilled, bonded, euphoric, and over the moon about having had a healthy baby, and this can actually create a sense of inadequacy and stress in new moms that they somehow aren't right. This stress then creates a vicious cycle of elevated corticosterone that reinforces the tendencies toward symptoms of PPD, which makes the depressive feelings even worse. Mothers experiencing postpartum psychosis may be at risk of physically abusing or even killing their infants.

In some cultures, there are customs to protect and nourish new mothers during their first postpartum weeks and months. For example, it is a Chinese custom for new mothers to "do the month" after giving birth. They do not bathe or leave the home; they are taken care of by their own mothers and are isolated to spend time in peace with their newborns and their family. Although there are many stipulations and regulations about what a postparturient mother can and cannot do at this time, the bottom line is that she is taken care of while she rests. A similar custom exists in Nepal, where women remain in their maternal homes and are secluded from general life and society for up to two months. Mothers are cared for and even fed a specialized diet for purification and milk production. Again, the main point is that the mothers themselves are taken care of, so that they can use their own energy to heal and focus on the baby. Perhaps the most important aspect of customs like these is that new moms are allowed to be removed from the world, and all of its extenuating stressful circumstances. Limiting the level of external stress on new moms is a critical point that many of us in the Western world seem to have missed.

Since this is primarily a book about animal moms, the burning question here is whether any of them are vulnerable to the effects of PPD. The answer is, of course, yes. However, it's impossible to simply ask another primate mom to explain how they are feeling and whether they may be depressed. What we can do is infer from behaviors and examine levels of stress hormones to get an idea of how other moms handle difficult postpartum situations. For many primates, their unfortunate predicament of being in captivity is something that causes them a great deal of stress. This can manifest in bouts of depression and result in child abuse. The pathology of depressive behaviors observed in female olive baboons, for example, is astonishingly similar to those experienced by human moms. Depressive baboon-mom behaviors include rocking themselves, pulling and plucking out their hair, and self-directed scratching—even to the point of skin-surface injury. Scratching is observed in several primates during periods of high anxiety, and is possibly an unconscious behavior that results from stress. In both captive and wild populations of macaques, the scratching behavior of females increases during the breeding season, and also during periods of group stress or danger. Whether in the wild or in captivity, the process of giving birth can be a stressful experience for females, and this stress is often quantified through monitoring the prevalence of self-scratching. Individual baboons that exhibit signs of stress during pregnancy or experience stressful conditions or circumstances during pregnancy are more likely to have their stress symptoms exacerbated after giving birth. This indicates that certain individuals, baboons or humans, are more likely than others to experience symptoms of postpartum depression.

Postpartally depressed macaque females exhibit symptoms including a slumped, broken posture and a low level of response to environmental stimuli. In a behavioral study that compared the maternal behavior of depressed mothers to that of "normal" mothers, biologists found that depressed mothers were more likely to hold their infants for longer periods of time but to breastfeed them less. Those infants of depressed mothers who were held for much longer periods grew into children and then adults who exhibited greater fear responses and were less able to cope with stress. High cortisol levels are associated with this predicament, suggesting that having a depressed mother can lead to long-term consequences in offspring. Non-depressed macaque moms spend greater periods of time engaging with their infants by grooming and playing, which is associated with the formation of social skills in the offspring. This is another area where the infants of depressed mothers lose out on the acquisition of important social and physical experiences.

It's perhaps implicitly obvious that depressed mothers, human or otherwise, are not in a psychological state to be as "good" as mothers who are not depressed. This is a negative and self-perpetuating cycle, as mothers can observe others around them who are potentially either doing it better or enjoying it more. In addition to having immediate negative effects on the mother, PPD can have negative impacts on the relationship between the mother and her new offspring, as well as the physical and social development of the latter. For any primate female, this must reflect poorly on self-esteem, which feeds into a greater level of depression and then an even lower level of maternal care. As mentioned before, most of the 300 primate species exhibit *exclusive* maternal care of offspring, meaning that dad doesn't provide direct care to infants. This is a massive responsibility, especially for the many species of Old World monkeys and apes that not only birth and exclusively care for their newborns, but also feed them breastmilk for a year or more. That there is no help for depressed primate moms (in the form of pharmaceuticals, anyway) means that they either raise infants in the best way that they can, or seek support from female conspecifics.

For mothers who are not at risk for developing depressive symptoms, the initial phase of infant care has common themes. Moms groom, feed, carry, and protect their babies for their first weeks of life. As they become a little older, mothers generally encourage a higher level of independence in their offspring, although they remain close at hand. Great ape moms play with their growing infants and share food with them as they begin to source nourishment other than breast milk. For several months, it's mom who makes sure that she's always near her infant; however, a shift eventually takes place in which it becomes the infant's responsibility to make sure not to wander too far from mom. Many primate moms will frequently break contact with their infants to encourage them to seek their independence. The frequency of this kind of maternal rejection increases as youngsters get closer and closer to weaning age, and this behavior is distinct from the rejection experienced by infants whose mothers are afflicted with depression or baby blues.

Maternal characteristics of protectiveness, warmth, and rejection are widely used to describe how a mother's individuality reflects on her parenting (as described above). These terms can be extended to apply to primate moms with depressive tendencies as well. Mothers who do *not* experience symptoms of depression are generally protective and not-rejecting of their infants for the initial phase of their lives, until it's time for them to become more independent, even if this results in pushback or tantrums. When there are negative impacts of psychopathology, macaque mothers tend to exhibit high levels of both protection and rejection (as mentioned above)—which undoubtedly has both immediate and long-term impacts on infant development.

When mothers are particularly afflicted with depression or psychosis, there is potential for them to become abusive toward their infants. Abuse of this kind has been mainly studied in macaques (both pig-tailed and rhesus), and the phenomenon is predictable in a small portion of mothers from both wild and captive populations. In large macaque populations the instance of child abuse is 5 to 10%.

Abusive macaque moms exhibit behaviors like dragging, pulling by the tail, throwing or pinning their infants to the ground, or stepping or sitting on them. The infant abuse occurs during the first month of infant life, and sometimes as early as the first day! This is clearly a time when the babies are the most passive and vulnerable, which means that they are likely not behaving in a way that promotes the occurrence of their own abuse. Effects of abuse range by the degree that an infant endures—from minor scratches and surface wounds to death. There are a few general characteristics of abusing primate mothers, most importantly that in addition to their abusive behaviors they also exhibit a full suite of competent behaviors too. In other words, in between short bouts of abuse, these mothers engage in normal and appropriate maternal behavior.

Depending on the severity of their abuse, mothers may exhibit normal maternal behavior most of the time and only abuse their infants in rare instances, or they may be mostly abusive in their interactions with infants, or anywhere in between. They are likely to be controlling of their infants, often keeping them in very close contact and scolding them when the infants attempt to break away. Maternal abuse of this kind may reflect inappropriate attempts of the mother to control the behavior of their infants. This kind of overprotective smothering inhibits the natural curiosity of infants, and they often become very clingy to their mothers as a result. Clinginess can also result from infants learning that whenever they stray from their mom they will suffer the consequences, and so after a while they simply stop trying. Abused infants begin to play at a much later age than non-abused ones, and when they do begin to play they do so much less frequently.

One of the most compelling aspects of maternal abuse in macaques is that it is repeated by mothers with successive offspring. This indicates that there is some kind of psychological and/or physiological difference between abusive versus non-abusive mothers. Indeed, mildly abused infants have elevated baseline cortisol concentrations long after the abuse has ceased, indicating that the effects of early physical abuse are long-term and likely irreversible. In addition, females are much more likely to be abusive toward their offspring if they themselves were abused as infants. In this way, the cycle of abuse tends to include a component of continuity along maternal lines. Sisters who were both abused have a greater chance of abusing their own offspring, which means that cousins will both be abused and become abusers and so on. Abused individuals also engage in the same type of abuse when they themselves become mothers. If infants are predominantly roughly groomed or bitten by their mothers or dragged or pushed by their mothers, then this is the kind of abuse that they inflict on their own offspring. This results in groups of related females that all abuse their infants in similar ways.

Experiments done on pigtail and rhesus macaques have indicated that the effects of abuse through generations are learned rather than genetic. Biologists have conducted many cross-fostering experiments, in which offspring of abusing mothers are switched immediately after birth to a non-abusing mom and vice versa. It's certainly probable that early abuse results in long-term and irreversible changes to neurodevelopment and the emotion processing centers in the brain such that abused infants are more likely to inflict the same treatment on their own offspring in the future.

Stressful conditions increase the probability that abuse will occur. In fact, the most common determinant of maternal abuse in macaques is stress from the social environment. Stress can result from harassment, lack of social support, low rank, infant kidnapping (which is more commonly experienced by low-ranking mothers), crowded conditions, group conflict, environmental perturbation, or predation. In captivity, the conditions under which primates are housed are often inadequate, which means that infant abuse rates are much higher in these populations. Keep in mind that in experimental scenarios where stress was induced, researchers have demonstrated that non-abusive moms do not become abusive. Not all mothers inflict abuse in reaction to stress, and not all moms who are in captivity are destined to become abusers. There are mothers who are already predisposed to abuse, and these individuals are more likely to engage in abuse when they are experiencing stress or captivity. Whether it's a physiological issue in the brain that regulates emotion, or whether certain individuals are more vulnerable to the effects of stress, the bottom line is that environmental factors act in conjunction with individual predisposition to provoke abuse in macaques and likely most other primates too.

Abusive macaque mothers continue their tirade until the infants are approximately three months old, when they gain physical and nutritional independence from their moms. This is unless the abuse has been so severe that the infant is unable to wean or has died. It's important to make a distinction between maternal abuse and maternal neglect. As mentioned earlier, extreme neglect or outright abandonment is most often perpetrated by young first-time moms. Biologists have not observed both abuse and neglect from the same moms—those who abuse do not neglect, and those who neglect do not abuse.

An important part of the abuse story that isn't well understood at this point is the long-term effects of abuse on the physiology and behavior of the abused infants. We know that abused infants are more likely to become abusers later in life, and this is due to some kind of developmental interference that happened while they were being abused. In all mammals, there is one unique feature that is a conserved aspect of our neurobiology—an intense, lasting bond with our primary caregiver (mom). This bond is established regardless of whether abuse occurs. In other words, a lasting bond is created with the mother even if she is mistreating the infant. There are direct modifications to infant cortical activity early in life that are associated with this bond, as well as the potential for alterations in epigenetic processes and activities of the HPA axis (hypothalamic-pituitary-adrenal axis, a major neuroendocrine system that regulates many physiological processes, notably those associated with stress). Infants with maltreating mothers often have increased cortisol levels that last well into adulthood, and there is initial evidence that negative early caregiving is remembered by molecular epigenetic markers as well.

TOP: Four host eggs from a hooded crow and two parasite cuckoo eggs.
CENTER: Two cuckoo (parasite) nestlings day 6, hooded crow (host) nestling day 0.
BOTTOM: Cuckoo (parasite) nestling day 19, hooded crow (host) nestling day 13.
All photographs courtesy of Yoram Shpirer.

ABOVE AND OPPOSITE: Rhesus macaque mothers and babies nursing peacefully.
Photographs by Kathy West/Kathy West Studios copyright © 2015.

ABOVE AND OPPOSITE: Like mother like child. Beautiful chimpanzee moments in Uganda. *Photographs courtesy of Adriana Lowe.*

A zebra mother and her child out on the plains. *Photograph by Kathy West/Kathy West Studios copyright © 2013.*

A mother weddell seal and her pup cuddle in the cold. *Photograph courtesy of Michelle LaRue.*

ABOVE AND OPPOSITE: A brown bat and her three-day-old pup. *Photographs courtesy of Brock Fenton.*

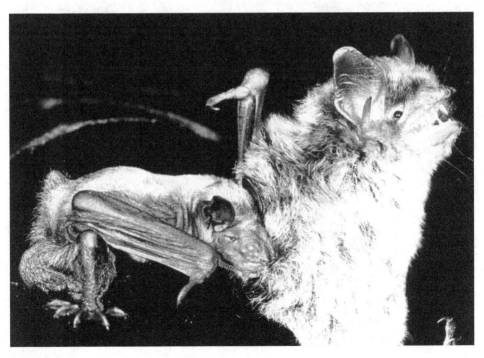

ABOVE: An elephant mother and her calf with females from their group. BELOW: A hippo mother and her calf in Uganda. *Both photographs courtesy of John King II.*

A lioness and her cubs on the Serengeti. *Photograph courtesy of John King II.*

A cheetah mother and her cubs on the Maasai Mara National Reserve. *Photograph courtesy of John King II.*

ABOVE: An Orca mother teaches her calves to hunt. BELOW: A South American sea lion mother and her pup. *Both photographs courtesy of John King II.*

A red leaf monkey and baby in Danum Valley, Malaysia. *Photograph courtesy of Unding Jami.*

ABOVE: A lowland gorilla mom and her baby. BELOW: A Rwandan mountain gorilla and her one-week-old newborn. *Both photographs courtesy of Joe Chowaniec.*

Happy *Homo sapiens* mom and her offspring. *Photographs courtesy of Sarah Sovereign.*

10

MAMMAL MOMS AND MILK

All mammals have mammary glands, and in females, these glands have evolved to produce milk. Milk is so much more than simply a food source for newborn offspring. In addition to containing the most appropriate combination of fats, proteins, and sugars to sustain species-specific infant growth, milk contains innumerable maternally-derived signals that contribute to many aspects of juvenile growth and development—hormones, immunological factors, growth factors, and so much more.

During the course of evolution, around 320 million years ago, amniotes (egg-laying animals) split into two groups: sauropsids, which went on to become modern-day reptiles and birds; and synapsids, which led to mammal-like reptiles, and eventually gave rise to all extant mammals. Approximately 166 to 220 million years ago, there was another split within this group, when the lineage that gave rise to modern-day monotremes split off from the lineages that gave rise to modern day marsupials and placental mammals. The marsupials and placental mammals split off from one another approximately 140 million years ago. So essentially there are three groups of modern-day mammals, two of which are much more closely related than the third.

The distantly related group contains the monotremes—or egg laying mammals—which includes only three species: short-beaked and long-beaked echidnas and the platypus. The monotremes are a fascinating group of animals for many reasons, not the least of which is their complex myriad of reptilian, bird-like, and mammalian characteristics. Monotremes lay eggs, have fur, and females produce milk for their hatchlings.

Monotreme milk is high in fat and nutrients; it provides all the food material that the young will ingest until they are large enough to leave the burrow and feed on their own. Echidna pups are housed in a small pouch on mom's abdomen, but platypus pups are located on the ground for the duration of their tenure in the burrow. This means that they are likely to be exposed to environmental bacteria and pathogens from a very young age. Special proteins have been isolated from monotreme milk that have antibacterial functions and aid in immune development. Monotreme Lactation Protein (MLP) is present in all monotremes and confers immune defense to otherwise defenseless neonates. Burrows are laden with opportunistic pathogens, bacteria, and various fungi and viruses which could all have negative impacts on early developing newborns; MLP and other bioactive compounds have evolved to stoke their immunological fires. Not only is milk the first and only source of nutrition for these pups, but it also plays a key role in the transfer of immune compounds, leading to direct development of the immune system while conferring protection.

Newborn marsupials such as kangaroos, wallabies, and Tasmanian devils closely resemble newly hatched monotremes. Marsupials are born at approximately the same developmental stage as what would be for a human fetus around the twelve-week mark. In other words, a whole lot of further development must happen (either in utero for placental mammals or within the pouch for marsupials) before these tiny newborns are actually ready for the world. This highly underdeveloped stage is when both monotreme and marsupial pups begin to feed via lactation; however, the modes by which they feed are quite different. The differences are a testament to the evolutionary history of both groups. Since mammary glands are a shared feature of both monotremes and the other two mammalian clades, we can safely assume that these structures evolved prior to their split from each other.

Mammary glands and milk represent two evolutionary anomalies in the animal kingdom. They are extremely important, present in all mammals, and there are no known intermediate structures aiding their function. Mammary glands are organs comprised of a series of complex ducts with secretory epithelial cells called lactocytes. These cells secrete milk into alveoli—which then expel milk via contraction. This kind of tissue, structure, and function, is unique to all mammals. In marsupials and eutherians, the ductal network comprising the mammary gland concentrates into a thickening of the epithelium where the milk is secreted, that is, through the nipples or teats. Monotremes do not have nipples, but specialized mammary patches where ducts are clustered and newborns can lick and suck to retrieve milk directly from the abdominal surface. The lactation experience is very different then between the monotreme young—who crawl around on mom's abdomen and obtain milk—and marsupial young, who crawl into mom's pouch and latch onto a teat for several months.

Marsupials have one of the most sophisticated lactation programs in the animal kingdom. Since newborns are highly underdeveloped, they remain permanently affixed to a milk-teat within mom's pouch for an extended amount of time. In addition to nutritional and immune-boosting compounds, marsupial milk also plays a role in directing development of specific organs like the gut. As the newborns reach different stages of development, the composition of the milk shifts to accommodate them. Unlike the milk of placental mammals, which remains fairly constant in composition after the initial colostrum, the milk of marsupials is a continuously changing matrix of what the offspring requires at any given time. As mentioned before, the mammary glands of marsupials are each controlled independently, which allows mom to provide dilute, gentle milk to a newborn from one teat at the same time that she is providing concentrated fat and protein milk to an older offspring that comes and goes from the pouch. It's as though the mother's body is a fine-tuned milk machine that serves up the exact concoction for maximum digestion and development to whomever is drinking it.

Although the specialized developmental strategies of monotremes and marsupials make them fascinating subjects for the study of mammalian motherhood they only comprise around 10% of mammalian species. The other 90% of mammals are Eutherians, or placental mammals—the group to which our own species belongs.

As mentioned previously, milk is far more than just a food source. It is essentially a species-specific solution to the problem of keeping newborns fully fed and appropriately developing. It is highly digestible, variably concentrated, and nutritionally balanced. The fact that mammalian moms are solely capable of producing the entire diet of their neonates is what allowed for the evolutionary explosion of novel habitat exploitation and diversification of developmental strategies in general. In other words, milk afforded mammals the freedom to evolve into a myriad of habitats and to become ubiquitous across the world. Many adult mammals have diets that would be impossible for their newborn offspring to digest (imagine a newborn human sitting down to a steak dinner) and milk provides the answer to the development of healthy offspring with fully developed digestive and immune systems. Lactation allows moms to collect nutrients and deposit them within their own tissues at one time and place, and then to mobilize these nutrients for their offspring at some other time and place.

Many of the compounds found in mammalian milk are not found elsewhere in nature. There are unique milk proteins called caseins (alpha, beta, and gamma caseins) and whey proteins, which provide amino acids to infants for growth and immunological provisioning against invading pathogens; they also play a role in fat digestion and uptake of nutrients from other milk components. Milk proteins play other physiological roles that scientists have yet to fully understand. For example, there are over 100 proteins in human milk, each serving its own biological purposes. Minerals are another important component: key minerals provided are calcium, phosphorous, sodium, magnesium, and potassium. There are also milk fat globules (MFGs) which are essentially membrane-enclosed lipid drops secreted as unique structures, and these are an evolutionary phenomenon of mammalian lactation. Milk fats provide support for metabolic energy as well as immune and developmental signaling. For example, large-brained animals exhibit a heavy requirement for polyunsaturated fatty acids for normal brain development. Milk sugars are another key component, and the main sugar in placental mammals is lactose, whereas in monotremes and marsupials, the main sugars are oligosaccharides.

In addition to these nutritional components of milk, there are hormones—such as insulin-like growth factor, glucocorticoids, leptin, and adiponectin—that can act individually or synergistically to regulate infant growth, physiology, and behavior. Bioactive molecules, growth factors, and enzymes are other milk components whose specific roles are being investigated but are likely impactful on things like infant growth, behavior, metabolism, and even sexual maturation.

Thinking about it in another way, milk pretty much picks up where the placenta left off in terms of how a mammalian infant is provisioned by its mom. The most energetically demanding time for infants, the period just after birth, coincides with the transition of physiological transfer of nutrients being very efficient via the placenta to being much less efficient via lactation. Where biochemical messages were once transferred between mom and baby through intimate placental connections, such messages must now be transferred through the less-immediate lactation.

The type of placenta in which the infant develops has an impact on several factors present in early milk, or colostrum. Colostrum tends to be more protein-rich and lipid-poor than later milk, and it is also higher in immunoglobulins and other innate antimicrobial factors. The level of dependency that a newborn has on colostrum consumption is highly associated with the specific type of placenta in which the neonate developed. In cows, goats, horses, and deer that have epitheliochorial placentae (non-invasive ones that keep mom's tissues separate from baby's), the transfer of immunoglobulins from mom to offspring during pregnancy is impossible. This makes the ingestion of colostrum and associated immunoglobulins and other immune factors absolutely *essential* for newborn survival. Carnivores that developed with an endotheliochorial placenta may also have had restricted gestational access to immune factors. On the other hand, baby mammals like primates and humans that have developed with hemochorial placentae, in which there is major contact between the tissues of mother and offspring, have already received such immune factors during their gestation, making the ingestion of immune factors in colostrum less important.

The duration of colostrum ingestion is an aspect of lactation that varies tremendously between mammalian species. Ursids such as giant pandas give birth to the most underdeveloped offspring among the placental mammals. For these animals, and for many bear species, there is a lengthy colostrum phase to adequately provision the helpless newborn. Colostrum provision in the giant panda lasts for thirty days, which is long—especially in comparison to the 24-hour colostrum provisioning in gray seals or the one to two days that humans get. The gut microbiome of newborn pandas undergoes almost a complete changeover in their first three to four weeks of life, which is a massive change that represents an extensive period of maternal investment. Pandas have an adult diet that is exclusively comprised of bamboo, so newborns require extensive seeding of the "correct" bacteria in their guts to be able to effectively digest the vegetation.

Something similar happens in a marsupial—the koala—although it's not colostrum provisioning. Mother koalas produce a substance called *pap*, which is ingested by their newborns when they first begin to emerge from her pouch. Pap is a euphemism for "feces." Baby koalas ingest their mother's feces as newborns to inoculate their gut microbiomes with the correct tannin-busting bacteria that will eventually be required when the babies start to eat their diet of eucalyptus. Pap has a higher water content and pH than mom's normal feces, and it contains high levels of the desirable bacteria. So, what colostrum cannot accomplish, a little poop apparently will.

Once the colostrum has flowed and then ceased, milk production begins, and lactation is the most energy-consuming undertaking of any mammal. More than gestation, lactation represents a sustained and substantial responsibility that is unlike anything else. The immense task of lactation is accomplished in so many diverse ways that an entire book could be written solely on the nuances of its variability across species (and in fact several books have been written to this effect), but there are some broad generalizations that can be made and a few unique examples of the lactation process are outlined here.

Biologists have coined a few terms to describe the use of resources by mothers to gain reproductive success. "Capitally breeding" mothers are those who save up resources as fat stores in their own bodies to be able to provision their young regardless of the immediate environmental conditions. Think about a pacific walrus mom packing on the pounds for her calf, who will suckle relentlessly for its first six months of life, tripling its initial body weight to the tune of 350 pounds from the milk mom creates from her fat reserves. Another classic example is the pregnant blue whale, which gains massive blubber stores of up to 99,000 pounds from foraging and feeding in temperate oceans. Moms give birth after migrating to more tropical oceans and provide extensive milk provisioning to their calves for a six-month period. Both the walrus and the blue whale mothers either fast or eat very little during their lactational phases, which means that they are dependent on securing enough resources to feed their offspring *before* the offspring has been born or even conceived. Blue whale moms produce 485 pounds of milk every day for their newborns— a colossal amount of milk production for one individual. Not all capitally breeding moms are aquatic, although if you're going to gain a few hundred—or thousand—extra kilograms of fat, being in the water is a pretty good idea. Some bears that have packed on the pounds prior to the onset of hibernation are able to give birth and lactate their cubs in their dens for a few months before emerging into the outside world in search of food.

Becoming extra-heavy in the name of maternal provisioning works well for species who can hibernate or who live in water, but it's simply not an option for many other mammalian moms whose ecology won't allow for it. Take bat moms for example, who need to maintain their ability to fly while lactating. They must therefore have a means by which to continuously find resources and provide milk for her offspring rather than bulk up and then fast during lactation. In addition, baby bats need to reach a critical threshold of development before they have any possibility of feeding themselves. Unlike other mammalian babies who can eventually crawl or walk to begin provisioning themselves with supplemental foods, bats must be able to fly, and their bones and muscles require high levels of protein and mineral support before they have sufficient strength to be able to accomplish this. This combination of a high need for lactational support but a low capacity for moms to store milk or large amounts of fat mean that bats are better classified as "income breeders." Income breeding mothers must spend time feeding themselves and their offspring, and they rely on their surroundings to provide food sources for nourishment.

While there are many species that can be considered capital *or* income breeders, there are many others that utilize a myriad of techniques depending on the immediate conditions. Humans and other primates are examples of animals that utilize both capital and income breeding strategies for lactation. Human females may accumulate fat stores during pregnancy (all in the name of providing resources for the newborn of course) that can be mobilized during lactation, but we also increase our dietary intake and restrict other behaviors that could needlessly drain energy. The truth is that since there is so much diversity not just between different mammalian species but also within members of the same species, and again within individual moms in different reproductive bouts, there is simply no straightforward formula for how a female may best utilize available resources for lactation.

Another critical factor to be considered is the length of time a female provides her offspring with milk. For some females, like Asian and African elephants, the lactation period is extremely long—up to eight years after a gestation period of nearly two years. Elephant calves nurse immediately after being born and receive colostrum for a short period of time before mothers produce mature milk. Baby elephants are exclusively dependent on mom's milk for up to six months after birth; however, they then enter a stage of lactation in combination with self-feeding which can last up to eight years. It's no surprise then that the milk provided by mommy elephants is dilute in nature. Generally, a female elephant will continue to nurse her offspring until she gives birth to another one, which is at least two years later but can be up to eight.

Humans and many other primate species also have a very long period of lactation, which generally means that the milks they produce are more dilute. Being able to produce a more dilute and thus cheaper milk over a longer time is advantageous for moms who may experience a disruption to their own resource base for short periods, or who feed their offspring on demand. Since they aren't putting all their available energy into lactation at any time, there is less fluctuation in milk quality and quantity offered to the offspring.

Such long lactations with high quantities of dilute milk are in direct contrast to mammalian species like hooded fur seals, who have among the shortest lactational periods observed. Hooded fur seals provide an incredibly dense, fatty milk to their pups for four days after giving birth. The milk produced is a whopping 61% fat, compared to the negligible 0.2% of fat in species such as the black rhinoceros, and the recipient newborn hooded seals gain up to fifteen pounds of fat on each day of their lactational phase. Essentially, it's four continuous days of consuming super high-fat, high-energy food that is meant to provide substantial stores to the newborn. The extremely cold polar climates mean that extra blubber is an essential layer of insulation for the newborns while they learn to fish and hunt for themselves.

The lactational habits of pinnipeds (walruses, seals, and sea lions) are actually very interesting. Three families make up the pinniped group, and these are odobenids (walruses), the phocids (seals), and the otariids (fur seals and sea lions). Walruses have the lowest reproductive rate of any pinniped species. Calves generally accompany mothers from birth, and they nurse on demand. The lactation period is approximately two years, and mothers teach their offspring how to forage on their own starting after the first six months. In contrast to the odobenids, many phocid species, like the hooded seal above, have very short yet powerful lactational periods in which moms provide large amounts of energy-rich milk and pups gain a large amount of weight quickly. Hooded seal pups increase their birth mass by 100% during their first four days of life, gray seal pups realize an average increase in mass of 236% in their eighteen-day lactation, and northern elephant seal pups become 300% larger over a lactation period of twenty-six days.

Moms of these phocid species lose a lot of their own body fat stores in a very short amount of time, and they generally do not eat during lactation. They haul themselves out onto the beach or onto sea ice to give birth and provide milk to their pups. The large bodies of many phocid seal moms allow them to gain significant fat stores prior to giving birth to pups, appropriate since they do not go back into the water until their lactational phase is over. Hence this strategy is called "fasting lactation." Weaning in phocid species is abrupt. Pups sink or swim (quite literally) with the provisions accumulated during lactation to survive the transition to nutritional independence. A short fasting period happens for the newborns once they are weaned but before they can hunt and forage on their own. They must learn how to provide for themselves during this fasting phase or they will simply starve to death. Once mom has finished providing milk, that's it.

Most phocid seals have very large bodies, greater than 330 pounds, that allow them to accumulate enough fat stores to fast during lactation. However, a few phocids (there are six small-bodied species) as well as the fur seals and sea lions (otariids) are smaller-bodied (less than 220 pounds), and must therefore provision themselves as well as their offspring during a longer lactational phase. The fact that otariids *cannot* fast for their entire provisioning phase has profound impacts on the species's lactational ecology.

Since mom must feed herself at the same time she provides milk for her offspring, she must leave her pups "parked" on land while she goes and gets food in the water. This is termed spatio-temporal lactation because while the act of lactation itself takes place on land; the nutrients that comprise the milk come from the ocean. Unlike many terrestrial newborns who can simply accompany their moms by following, such as elephants or other ungulates, the newborn otariids do not yet have the skills to manage underwater and they would undoubtedly make a mother's foraging excursion impossible if they were to tag along. Not only does infant parking mean that mom is away from her offspring for long periods of time, it also means that she requires a way of recognizing her own pups when she returns from foraging trips that can be as long as sixty days. The length of a mother's foraging trips has a lot to do with the availability of prey in the immediate environment. Trips can range from a few hours to many weeks. The pups will fast during the entire period of mom's absence, so they can easily starve to death if mom doesn't return in enough time with enough food or if she doesn't come back at all.

However, in most cases mom *does* return in time to her waiting pups—but a fairly substantial task awaits her, which is to find her own calf among the hundreds of visually similar infants in the rookery. And yet, natural selection has favored intricate and effective recognition mechanisms in both moms and pup otariids so that they find each other quickly. After all, providing care (especially expensive milk) to pups that aren't hers is a waste of a mother's energy, and a pup that solicits the wrong mother for milk may be putting itself in physical danger from her resistance.

In subantarctic fur seals, the first separation of mom and calf happens very soon after birth, at approximately two to eight days. It's critical for newborns to acquire the ability to recognize their own mom before she sets out on her first foraging journey. Unique characteristics of both mom and pup calls and specific aspects of their individual voices are utilized by both to find each other. Presumably, Antarctic fur seal mothers delay departure until they are certain that their infants will recognize them when they return—so the first foraging trip could be as early as two days after birth or as late as eight. Reunification seems a daunting task, but most mom and pup combos can find each other within a mere seven minutes of moms return to the breeding colony. Mothers rely on the frequency modulation of their own pups' calls to recognize them, whereas pups rely on the energy distribution of their mom's calls. Both mom and pup use a second stage olfactory cue to verify each other's identity once they have come close enough to each other to do so. The fact that recognition is so fine-tuned in many otariid species is a testament to its evolutionary importance. Suckling sites can change by up to thirty-two feet when pups are small (simply by the pups wriggling around in mom's absence) and even greater distances when pups are older and more mobile. Mom simply cannot therefore rely on spatial cues about where to find her pups.

Although an otariid mother can be away from her pups for many weeks at a time, her milk supply remains at the ready when she returns to her pups. Any breastfeeding human mom who has gone back to work knows that if you don't keep pumping your breasts, you'll quickly lose your milk supply, not to mention suffer the engorged and potentially infected ducts that one may experience from quitting abruptly. Indeed, for most mammals a prolonged separation from suckling initiates mammary gland involution and cessation of milk production. Humans, like phocid seals, require regular evacuation of their mammary glands to maintain normal milk supply. However, the same is not true for otariids, whose mammary glands remain fully functional even after long periods foraging at sea. In the case of the otariid cape fur seal mother, the lactation period is approximately ten months long, and it consists of a repeated cycle of long at-sea foraging trips of up to twenty-eight days at a time followed by short suckling periods of two to three days on shore. It takes only a few days for her hungry pups to entirely deplete her milk supply before she needs to hit the road again.

Lactation can also be affected by environmental characteristics. Access to appropriate food and water is critical to any *individual* lactating mom, as both the composition and volume of milk that a female produces depends on her own physiological state. However, there are also common themes in milk production at the *population* or *species* level that result from constant and predictable environmental conditions. For example, animals in arid environments produce milks that are more dilute, to provide offspring with greater hydration and to facilitate evaporative cooling. Moms that feed their offspring on demand, or whose offspring accompany them twenty-four hours during a long lactational phase, also have more dilute milks. Mammalian species with small body mass tend to produce milks that are more concentrated and energy dense than species with higher body mass, based on the reduced digestive capacity of a smaller gastrointestinal tract. Carnivorous species are thought to produce milks that are higher in fat, protein, and energy because animal matter typically contains more fat and protein than plant matter.

Reproductive output is another factor that plays a major role in the lactational process because a mom that gives birth to an entire litter of pups clearly needs to have a different strategy than a mom who gives birth to only one. There are many factors and generalizations that should be kept in mind when considering the parameters of lactation for any given species. There are also many factors to keep in mind when observing differences in the lactational output of different moms of the *same* species. It is this kind of difference that is perhaps the most fascinating when it comes to the consideration of motherhood in the animal kingdom. How do social and ecological factors work their way into milk production and distribution?

A great place to start in this regard is with the spotted hyena. Hyena societies are matriarchal, which means that females are the socially dominant sex. They live in groups of sexually mature females and their offspring. When young males reach sexual maturity they are kicked out of the group. Adult males are low on the social scale, they aren't generally permitted to socialize with the group, and often end up in small bachelor pods or as solitary individuals. There is one alpha female within a hyena clan, and although there are many reproductive females in the group, the alpha is, without question, highest on the social scale. Subordinate females and their offspring are socially submissive to both the dominant female and her offspring (see *Cooperative Creeds and Communal Cliques*). This hierarchical structure sets the scene for complex lactation dynamics. Pups feed exclusively on mom's milk for approximately six months, and they are generally weaned between twelve and twenty months. The cost of lactation for spotted hyenas is generally high to begin with because their main prey species are migratory ungulates like gazelles and springbok. When prey abundance is high, mothers can forage locally and remain in close contact with their cubs in a common den. However, when their prey is absent from the immediate territory, mothers must embark on long commutes to find them. Such trips can encompass hundreds of miles, and mothers are not only absent for long periods, but they are also exhausted upon their return. Their plight is similar to that of the female otariid seals who travel various distances to find enough food to build up their milk supply.

Whether a hyena mother has to go on a long foraging trip to produce enough milk for her cubs is entirely dependent on rank. When food is scarce, high-ranking mothers claim the nearby hunting territory, forcing the lower ranking females to forage farther from the communal den. Pups are sometimes forced to fast for up to nine days while mom is away searching for food. The long-distance foraging trips are stressful for low-ranking moms, as fecal glucocorticoid levels have demonstrated. Although the immediate effects of milk-glucocorticoid ingestion have not yet been researched in hyenas, there are measurable and predictable impacts of milk-glucocorticoids on infant rhesus macaques (see discussion below). Not surprisingly, the amount of milk transferred to hyena cubs increases with decreasing social status—which reiterates that low status moms are more often away from the den and therefore provide larger meals to their pups upon their return. The quality of the milk also increases with decreased feeding bouts, so that low-ranking moms provide higher amounts of high-quality milk, just at lower frequency feedings than high-ranking moms. Growth studies have confirmed that pups receiving more frequent feedings from high-ranking moms have higher growth rates than those who are forced to fast and then have more nutritious meals.

Another unfortunate predicament for hyena mothers is susceptibility to pathogen infection. Mothers that continuously make massive commutes to feed their offspring not only experience a greater level of stress, they have a lower proportion of available energy for body maintenance and immunity. Generally, as hyena social status decreases, the incidence of parasitic or pathogenic infection increases. This is a common issue for lactating mammals across the world, humans included. When one is performing the energy intense service of lactation, there is simply less energy to be allocated to other metabolic needs.

The long fasting periods between feedings experienced by pups in the litters of low-ranking moms can also affect their own relationships with siblings. Hyena moms usually give birth to twins, and from birth these pups have strong jaw muscles and sharp teeth. Siblicide is common in hyena litters, as the most aggressive pups are likely to obtain the greatest share of mom's milk. Indeed, hyena moms often allow an aggressive pup to edge its sibling out of a meal time and time again until it starves to death. In the hyena world, there is simply no room for a weakling. If mom is away foraging for long periods, pups are more likely to increase their aggression toward each other to be the one who feeds first upon her return. Indeed, dominant siblings of low-ranking moms achieve higher amounts of milk consumption than their weaker counterparts. Although mom would ideally aim to maximize her biological fitness by keeping both pups alive, if resources are scarce she may be better off to allow the dominant sibling to starve the weaker one.

Although spotted hyena moms allow their cubs to duke it out for access to her teats, in other species mom tailors her milk depending on which of her offspring is drinking it. Many mammals and primates tend to provision male offspring with milk that has a higher fat content. Females receive a lower-fat and thus lower quality version, but they may receive larger amounts. In many primates with polygynous mating strategies (one male several females) and sexual dimorphism (males bigger than females), mothers provide their male offspring with additional food or milk with a fattier composition because they have a much larger cost to being small than females do. A male's size will have a massive impact on his future social and reproductive success, and so it makes a lot of sense for a mother to make sure her sons have the best chance to get as large as possible. On the other hand, many moms provision their daughters with extra calcium stores from their milk. Females generally have an accelerated rate of skeletal calcification than males do, and so mothers can provide milk that facilitates the process. In these ways, mothers tailor their milk for each infant.

The complexities of breastmilk composition are illustrated by work done on rhesus monkeys that shows offspring sex to be just one factor that is involved. Although most work has been done on its nutritive components, scientists are just now starting to unravel the additional resources (like bioactive molecules) provided in breast-milk, and the fact that many of these factors are as fluid as the milk itself. Bioactive molecules are those that have an active function in the body of the ingesting animal—in this case the infant. For example, in rhesus monkey milk there are bioactive cytokines that bind to receptors in the infant gut. Instead of being digested, these molecules have a direct role in the further development of the infant gut, and they are just one of a suite of bioactive factors that is being investigated. Rhesus macaque mothers have different combinations and concentrations of milk bioactives that are influenced by factors like social rank, mother's weight, and parity (how many babies a mother has had previously). The fact that these important messages are being transmitted to baby macaques based on both physiological and ecological conditions of individual mothers is something that is likely important in other primate species as well.

In addition to the varied concentrations of bioactives in mothers' milk, mothers can also feel varied levels of stress during lactation—resulting in varied concentrations of stress factors in milk. Primiparous moms in the animal kingdom—specifically rhesus macaques—are often still growing themselves, which puts them in a position of making tradeoff decisions for current and future reproduction. They must decide if they should invest less in the infant now and make themselves stronger for future reproduction or invest more in this infant because making a weak infant is going to be a hit to their biological fitness. In addition, first-time moms have proportionally larger infants to care for because they themselves are smaller than older moms, and they have fewer fat stores from which to draw energy for lactation. The ability to synthesize milk tends to increase with increased number of parities because of the cumulative effects on the architecture of the mammary glands, meaning again that moms with other babies are at a considerable advantage over their primiparous counterparts. What does all of this mean for primiparious moms? It means a great deal of stress.

Stress in first-time moms is elevated, and it's exhibited by increased levels of circulating glucocorticoids in both the bloodstream and in the breast milk. It's generally understood that prenatal exposure of infants to glucocorticoids while still in utero can result in low birth weight and a compromised level of immune function and neurodevelopment. Less is known about how these stress hormones may function when they are received in the breast milk, but studies of macaques has provided some insight. Primiparous moms with elevated levels of glucocorticoids (cortisol) in their breast milk tend to have infants with more nervous, less confident behavioral phenotypes than multiparous mothers. In other words, elevated glucocorticoids in breast milk is strongly associated with changes in infant temperament, and this is true for both sons and daughters. Higher cortisol concentrations in milk is associated with a greater infant weight gain over the long term. This seems counterintuitive upon first glance, but the more nervous, less confident infants do not generally engage in exploratory or playful activities. Instead, the bulk of their ingested energy is channeled directly to growth. In remaining "hidden" by their moms, the more nervous infants are indirectly saving all their energy to get big and strong. However, they are also potentially losing out on learning and social opportunities to be gained through interaction with others. An evolutionary coin-toss.

Indeed, behavioral boldness is associated with reproductive success across many animal species—and primates are no different.

This trend is especially true for males, who are often favored more by females if they are the bolder in nature. However, the strategy employed by the primiparous macaque moms implies that there is something to be gained by both male and female offspring by channeling as much energy into their own growth as possible. Perhaps, since offspring of primiparous moms are naturally at a disadvantage due to their physiological and social status, the generation of more nervous phenotyped babies that grow more over the long term is a tradeoff that allows for these infants to survive. If you don't survive to sexual maturity in the first place, there isn't much sense in being bold and confident as a baby.

11

OPPORTUNISTIC ORPHANS
AND MAGNANIMOUS MOMS

M oms can employ a number of strategies to ensure that their young are adequately fed during their first few months of life, and many of these involve providing help to other moms by sharing in the lactational responsibilities. Termed allonursing, we know it as wet-nursing—the act of providing milk to newborns that are not one's own—and it's very common in the mammalian world. To date, allonursing has been observed in sixty-eight different species across all major mammalian groups, which indicates that it is adaptive for a wide variety of mammals in a wide variety of situations. In addition to the obvious nutritional benefits of allonursing, infants suckling from multiple females (in any group situation) obtain the benefits of increased diversity in immune factors transferred from each lactating female.

Since the act of milk sharing between females is so common, a variety of hypotheses attempt to explain its existence. It's likely that milk sharing has independently developed in a number of mammals to suit the unique biological and ecological niches of various species, and there are many good reasons for doing so. Keep in mind that lactation is an exceedingly expensive process for females, requiring more energy than the actual process of gestation. In light of this, making the decision to share milk—an important and expensive resource—is not a trivial one.

Among the most supported hypotheses for doing so is the *kin selection hypothesis*, which simply states that mothers are more likely to share these expensive resources with the offspring of close relatives. One's overall inclusive fitness can be increased by providing sustenance to related newborns. This kind of action is observable in collective breeders like meerkats and mongooses, in which many of the females providing the allonursing service are doing so for the offspring of their dominant mother; hence feeding their sisters and brothers. Even non-reproductive females of several species spontaneously lactate and feed related young when the situations present.

The related *social benefit hypothesis* is limited to situations in which a significant gain in social standing can be derived from suckling someone else's offspring. In cooperatively breeding groups, there is often a discrete social hierarchy to which most members adhere. There are both singular (where only one mother reproduces and creates all the babies for a given group) or plural (where two or more females in a cooperative group are reproductive) cooperative breeding groups. Although in the plural groups many females may be reproductive, the social hierarchy still exists, and low-ranking females may still provide an allonursing service to the offspring of high-ranking females to gain the social benefit. It is unlikely that the high-ranking females return the favor. Instead, allonursing provides a way for low-ranking females to show their subordinate status and presumably gain the social favor of those higher on the social scale.

In other group-living situations, such of those of African and Indian elephants, more than one female may be reproductive without the existence of such a notable hierarchy. Such societies are matrilineal, with mothers, sisters, and offspring forming tight-knit groups. Instead of there being specific ranks, these units function as mini-close-knit families of females without the influence of their male partners who live in solitude or in small bachelor pods. Within these all-female groups, the newborns may feed from any of the lactating females, although the bulk of their nutrition comes from their own mothers. In addition to providing several modes by which the juveniles in the group can gain sustenance, allonursing each other's offspring also facilitates the formation of strong and lasting bonds between group members. Similar group-living scenarios are observed in many carnivorous species like lions, where females keep their young together and nurse reciprocally. There is usually a high level of relatedness between adult females in these groups, although they differ in composition from elephant groups in that males are part of the picture as well.

Another hypothesis for milk sharing is one of reciprocity between familiar mothers. In this case, there is little or no genetic relatedness between females; however, close bonds form between females who see each other all the time. Group living mammals and other species who reproduce communally or share a common roosting ground have the potential to become both friendly and trusting of the individuals in their immediate vicinity. When mothers give birth at a similar time and find themselves in parallel situations with respect to immediate childcare and foraging duties, they can work together to increase their success at both. This is called the *reciprocity hypothesis*, and it's highly supported in data from various group-living species.

In many species like house mice, females choose a specific female alloparenting partner. Such choice is not random; females carefully consider exactly which other female to pair with to create childcare dyads. Considerations include familiarity with each other, common living area, and similar stage of gestation. Once the partnerships are formed and the litters are born, females pool their pups in a common nest and, from this point forward, it is impossible for them to tell their own from those of their dyad partner. Females care for all the infants together and nurse them indiscriminately. This kind of motherhood sharing serves to increase the lifetime reproductive success of both females involved in it. However, the importance of selecting an adequate partner is not to be underestimated. Females must provide relatively equal care to their collective broods, and cheating must be identifiable and punished. Poor partner choice can result in total brood loss through infanticide or neglect if one female is left to provide milk to offspring of both litters and to maintain all other biological duties as well.

The notion of familiarity versus relatedness is an important one for several aspects of motherhood. Behavioral biologists are just scratching the surface of understanding the complicated nature of relationships in the animal kingdom, and there is immense diversity both within and between species. Common sense will tell you that human mothers vary greatly when it comes to our parenting decisions. We therefore tend to associate with mothers who have similar views about things like feeding, recreation, and the like. Although the house mice take co-parenting to the extreme, it's true that a lot of good comes from co-mothering with familiar individuals.

The allonursing discussion thus far has largely centered on females that give birth to litters. When there are many newborns to contend with at any given time, there is less individual attention for each. Allonursing is therefore more common in polytocous mammals, and it can be initiated according to the kin selection and reciprocal hypotheses outlined above, or it can also be due to a simple case of mistaken identity. First-time moms may be more likely to allonurse newborns that don't belong to them because they lack the experience and knowledge of multiparous or older moms. This situation is described by the *misdirected care hypothesis*, and it's seen in both monotocous and polytocous species. In polycotous species living in groups, the costs of making an allonursing mistake are not exceedingly high. The benefits of living in a group—reduced predation threat, high genetic diversity, and overall higher level of offspring survival—far outweigh the costs of occasionally getting it wrong.

Comparative modeling across all mammalian species shows that for a given mass, all females produce the same amount of offspring mass. In other words, a ten-pound monotocous newborn is equal to ten one-pound polytocous newborns. In terms of the level of care given by mom to each newborn, each individual in the polytocous species can expect 10% of the care whereas in the monotocous species the single newborn can expect 100% of the care. This highlights the discrepancy between the costs of allonursing in both systems. A polytocous mom is only giving 10% of her effort to each of ten pups (or whatever percentage, based on the number of pups she has), meaning that if she happens to accidentally allonurse another pup, it's not a huge loss for any one pup individually since each one was only getting 10% in the first place. However, if the monotocous mom accidentally allonursed an alien offspring, her own offspring would surely notice any kind of decrease from the 100% of mom's care that it was already getting. This makes the cost of misdirected care in monotocous species especially high, and we would therefore expect these females to quickly learn how to discern their own offspring from others.

In Steller sea lions, misdirected care in the form of allonursing is common among first-time moms, whereas older, multiparous moms are very rough and unforgiving of unrelated juveniles that attempt to suckle from them. More experienced moms shove or throw unrelated pups that are attempting to nurse, often biting and significantly injuring them in the process. The milk of Steller sea lions is particularly high in fat due to the high fat requirements of the pups to thrive in a northern ocean environment, and so females have clearly developed some very strict practices to ensure that their own offspring are the primary recipients of this rich resource.

In other monotocous mammals that breed in groups, there could be a certain tolerance of allonursing infants if all moms in the group are acting in a similar fashion (*kin selection hypothesis* or *reciprocity hypothesis*). If each individual mother is spending a similar amount of energy on unrelated pups, then one would expect this behavior to be sustained and promote both peaceful function and group cohesion. Indeed, this kind of situation is observed in some groups of giraffes, camels, and reindeer, although it is very difficult to distinguish maternal tolerance from potential milk theft.

The fact that some newborns can take milk from moms other than their own deserves further consideration. In these cases, it's important to discern whether the situation is best described by misdirected care, in which a mom has a hand in mistakenly identifying an allosuckling pup for her own pup, or whether it's a case of milk-stealing—a behavior of the pup that is outside of the awareness of the allonursing mom. Many mammalian newborns are much more capable than human newborns. They know whether they are being adequately fed—and if they are not, they are capable of sourcing out an alternative mode of nutrition. Milk-stealing is quite a common occurrence in ungulates like reindeer, camels, and giraffes, and offspring undertaking this risky behavior have a few common strategies.

First, milk-stealers generally try to suckle from a mom while her own offspring are suckling as well. This works well when there is more than one natural pup in a litter and the milk-stealer selects a position in the middle of the pack. Second, many milk-thieves will position themselves in an anti-parallel position with respect to the mom so that their back end is closest to the mom's front end. Mom is less likely to realize that the intruder is not her own if she isn't seeing the face. This anti-parallel strategy works especially well if used with the combined feeding method. A milk-stealer can be successful even if their stealth is not verified, as of course, it's difficult to conclude whether an infant is actually fooling the mom or whether the mom is simply tolerant of allonursing because of a mutual understanding with the moms in the group.

As mentioned earlier, there are potential high costs to milk-thieves in monotocous seals and sea lions, where multiparous females will aggressively and violently thwart allonursing attempts. In Steller sea lions and other closely related species, allonursing attempts are common and often unsuccessful unless the mother whose milk is being sought is asleep. Indeed, hungry pups will be very crafty by waiting until a suitable candidate is far away in dreamland and then stealing her milk with a reduced chance of being punished. At this point, the behavior is better described as a form of parasitism rather than allonursing, and it is a behavior that is entirely of the pup and not of the mom. Young sea lions and seals can make a living at milk-stealing, as one recent set of observations found that despite losing its mother when it was just weeks old, a young male pup was able to gain enough nutrition to wean with his cohort at a normal size, which occurs at approximately six months. He did this without being adopted by another mom; rather, he simply stole milk from sleeping mothers whenever it was available.

Pinnipeds like seals and sea lions breed in dense colonies where females haul themselves out onto land to give birth and care for their newborns. The dense colonies are a form of protection from predators and offspring tend to have a higher survival rate when reared in groups with other moms and pups. The orphan newborn just described survived without his mom because this type of communal breeding strategy facilitated multiple bouts of milk-stealing. However, we would not expect milk-theft to be an effective strategy at all in species that have a more solitary social system. In addition to pups that learn to "outsmart" unsuspecting allomothers in group living species, there is also the opportunity for young juveniles in such species to compete with each other for access to a mother or allomother. Seal pups will often push off a nursing colleague to gain greater access to the teat, potentially harming them in the process. All's fair in the natural world.

Not to be confused with the previously mentioned *misdirected care hypothesis* for allonursing, the *parenting hypothesis* posits that first time or moms who haven't given birth will allonurse unrelated offspring as a means by which to gain experience and skill for when the real thing occurs. This kind of behavior is observed in many primate species, such as Japanese macaques, in which both related and unrelated females will suckle newborns. If a female is pregnant for the first time and she's in her last few months before giving birth, she is likely to be providing milk to allonursing infants just because her body is able to make it. However, in many other cases, pre-birth females are not providing milk to newborns while they suckle; that could mean that the female is either "fooling" the newborns into suckling so that they can gain practice or that there is some kind of comfort and calming benefit afforded to juveniles by the suckling process. The *parenting hypothesis* could also apply to inexperienced females that spontaneously lactate without reproducing, or to females that have lost a pup or litter due to extenuating circumstances. In these latter cases, females can potentially gain valuable parenting experience while providing assets to unrelated newborns.

The likelihood of a young juvenile obtaining comfort via suckling is high throughout the mammalian class, and it is certainly demonstrated in our own species as well. This is termed the *soothing hypothesis*. In primates like tufted capuchins, allosuckling is generally used by non-lactating females to soothe fussing newborns. A similar trend has been shown in African elephants, where the main function of allosuckling is to provide comfort to stressed infants and is primarily performed by nulliparous females who are unlikely to have milk in their teats. Since the soothing function of suckling is common across mammals, allowing troubled infants to do so can have benefits that extend to the entire group. Stable social groups rely on the continuous efforts of members in many ways, and soothing infants is most certainly a way in which non-breeding females can maintain positivity and peace in group function.

Although there is support for all the aforementioned hypotheses, there remain other reasons for the existence of allonursing in the wild. Any human mother with an engorged breast will know that it's often a good idea to expel extra milk to avoid painful buildup or possible infection. In other mammals, when one's own offspring are satiated, or if a mom has had a smaller-than-usual litter and has milk left over, mothers will often seek out another candidate to feed to decrease their load. Most often such allonursing practices—described by the *milk evacuation hypothesis*—are peaceful and mutually beneficial. However, in at least a few species, such as elephant seals, females sometimes forcibly kidnap unrelated offspring to help them decrease their milk load. The fact that extra milk can usually be consumed by an allosuckler is helpful for a number of reasons apart from avoiding engorgement or infection. Female evening bats allonurse unrelated offspring prior to heading out on foraging bouts, which substantially lightens their load prior to having to fly great distances. Having an excess supply of liquid on board is a substantial detriment to one's flying ability, and bats are unique in the animal kingdom as the only flying mammals.

Many females continue to allosuckle after their own offspring have weaned as a means by which to lose excess body weight. For example, as previously discussed, several seal and sea lion species haul themselves out onto land to give birth and care for their newborns. Elephant seal females do not forage during the initial weeks following giving birth; they instead rely on stored fat reserves to carry them through. Before they can return to their pre-birth foraging habits in the deep sea, they need to lose all the extra fat reserves that prevent them from diving (i.e. their bodies remain too buoyant with excess fat). If their own pups have finished suckling, these females will continue to allosuckle other pups until they have depleted their fat resources.

12

OTHER MOTHERS:
ADOPTION AND ALLOMOTHERING

The copious instances of cooperative care that we observe in the animal kingdom often amounts to nothing more than moms simply lending a hand to other moms. Humans do it all the time, when they have their children's friends over for meals, playdates, or sleepovers—or even when they inadvertently keep an eye on all the children at the playground to make sure everyone's okay on that big slide. This is allomothering, and it's something that most if not all mothers provide to each other at some point. These aren't tasks that necessarily go a long way to contributing to a mom's ultimate demise or accumulation of biological fitness; they are easy day-to-day things that many mothers do all the time without even really thinking about it much. Such allomothering is so common in the animal kingdom that it would be an exercise of brutally boring proportions for me to describe it. Suffice it to say that animal moms of social species are generally good to each other and are almost always on both the giving and receiving end of some form of allomaternal care. The kinds of behaviors that are exchanged or provided are things like feeding, grooming, protection of infants, and carrying of infants. It's possible in some cases that moms exchange goods for services if one mother is out foraging or hunting while others are watching her offspring. And there are times when allomothers are required for more specific tasks. For example, if a human mother is going on a work trip or trying to have a date night, she'll have to make arrangements for alloparenting of the children while these things take place.

Nazca boobies are ground-nesting colonial seabirds that exhibit unique pseudomothering behavior. In any given year, there are reproductive females whose nests have failed. While busy parents are off foraging for food for their nestlings, which can take a long time, depending on the availability of resources in a specific year, these females, and sometimes males too, pay "visits" to unattended chicks. These visits can last for varied amounts of time, from one minute to over an hour. Newly hatched chicks and almost-fledged chicks are not visited by these pseudomoms because biological moms almost always guard new hatchlings, and chicks that are ready to leave the nest are large and strong enough to defend themselves. It's the chicks that are in the middle that receive the visits, they are the "goldilocks" age for vulnerability—too young to defend themselves but too old for their mothers to remain at their side. During pseudomom visits, chicks assume a submissive posture with their heads inclined forward and the dorsal surface of their beaks on the ground.

Some of these visits are friendly and peaceful. Females will stand beside or preen young chicks or, in some cases, even provide them with gifts of pebbles or feathers. Unfortunately, the bulk of the visits involve aggression by the visitor, who will scratch or bite chicks, leaving them with bloody wounds on their necks. These wounds aren't fatal, but they serve to attract the attention of blood-sucking ectoparasitic birds like finches or mockingbirds. These parasites feed from the cuts, which deepens them significantly—often leading to the death of the chicks. This strange behavior is very common in Nazca booby colonies. It's observed across different populations in different areas, and therefore cannot simply be dismissed as erratic behavior of a few females. Up to 25% of booby chicks die as a direct or indirect result of these unwanted visits.

The Nazca booby example is extreme, as such strange antics aren't generally practiced among mothers of vertebrate species. For the most part, things tick along without infants having to endure anything even close to the abuse suffered by the little Nazca booby chicks. Allomothers and adoptive mothers across the mammalian and primate worlds provide each other with various levels of camaraderie and support. However, there is a very real difference between allomothering and full-on adoption, in which a non-biological mother takes over complete care of an orphaned infant. The latter represents a much greater commitment and is much less common.

Biologically speaking, it doesn't make sense for a mother to provide an *abundance* of care to an unrelated offspring. There is no way for a female to increase her biological fitness by adopting an infant that doesn't carry her genetic code. So why are there instances of adoption in many animal groups? It turns out that there are several well-supported hypotheses about the existence of adoption that are dependent on aspects of the ecology of the species in question. Seeing as adoption is a massive commitment—much like the energetic commitment of allolactation—the hypotheses for its existence are along similar lines. Adoption in primates generally involves relatedness between mothers such that overall inclusive fitness is increased by doing so. This is called the *kin selection hypothesis* and could easily represent a scenario where a sister or aunt adopts an infant following the death of its mother.

Many detailed instances of primate adoption are recorded in scientific literature, but there is no consensus with respect to the specific conditions under which one would expect it to occur. This isn't particularly surprising, seeing as primates are exceedingly complicated creatures with a myriad of factors contributing to most aspects of their reproductive and social lives. To give examples of the variable array of conditions under which adoption may occur in primates, here are four case studies that are particularly compelling and illuminating.

The first is documentation of a "baby switch" between two woolly spider monkey females. Female A had a female baby, and while she held and nursed her own newborn, she also held an eight-day-old male infant that had been born to another female in the group, female B. A few days later, and for several subsequent observational trips, female A was seen exclusively with the male infant, while the newborn female infant was cared for by the mother of the male, female B. Both infants were well cared for and successfully weaned to independent ages—just by reciprocal mothers. It's difficult to judge whether this was a case of mistaken identity, or whether the two mothers were related, which could be explained at least part by kin selection.

In a second case study, a female capuchin monkey immediately adopted a three-day-old male infant that had been abandoned, despite the fact that she herself had a four-day-old infant at the time. She raised the male infant along with her own, and spent equal amounts of time in grooming, carrying, and cuddling both infants equally. The only observed difference in care between the capuchin mother's own female infant and her adopted male infant was that the male infant spent more time suckling. Suckling duration is a common difference between male and female primate infants—males often feed more in order to meet the challenging competitive needs that males face in later life. This makes it quite possible that the adoptive mother was simply provisioning her male child as a normal primate mother would—with no favoritism toward either her own biological infant or her adopted one.

Females who have recently given birth are naturally motivated to interact with infants and are much more likely than non-parturient females to adopt an unrelated baby. In this latter case study, both the infants survived to weaning age, showing that a female capuchin can successfully rear an adopted infant along with her own. The fact that this female also had older offspring that helped with carrying the two infants was also an important aspect of the success of this particular adoption. This kind of "additional" baby scenario has also been documented in both gorillas and macaques. Imagine for a moment, if you had just given birth to a baby, and you then found a newborn baby abandoned by your door. Assuming you are in the same position as the capuchin mother (that is, no police/hospital/911—just your own conscience), would you take the baby in? If there was nowhere else for that baby to go and your body was primed to deliver the appropriate goods, my guess is that you wouldn't let that little defenseless newborn die. Why should we assume that other female primates wouldn't do the same?

The specific capuchin case study just described was indeed a wonderful culmination of several social, environmental, physiological, and reproductive factors that resulted in the successful rearing of both the orphan and the biological infant. Unfortunately, the situation is not the same for the third case study, in which an orphaned infant was adopted by a mother in a group of colobus monkeys. The mother adopted an abandoned orphan and spent time raising it alongside of her own infant, much the same as the capuchin in the previous example. However, in this case her biological offspring subsequently died, possibly because the mother couldn't provide it with enough care after adopting the other infant. The abandoned infant persistently clung to and fed from his adoptive mom, which could certainly have contributed to the biological infant's demise. Once the biological baby died, the mother cared exclusively for the adoptee, who made it successfully to independence.

One last primate case study that is particularly interesting is a case of adoption between species. An abandoned infant marmoset was adopted by a group of capuchins on a biological reserve in Brazil. Two of the mothers in the group provided successive care to the adoptive marmoset, feeding and grooming it, or carrying it around. The most striking aspect of this adoption is the sheer size difference between the two species. Capuchins are generally 6.6 to 8.8 pounds whereas marmosets are just three-quarters to one pound—exceedingly tiny in comparison. The little marmoset was like an infant doll or pet that was being cared for by its adoptive moms. Another fascinating aspect of this situation is the fact that marmoset infants are normally weaned very quickly and are independent individuals much earlier than capuchin babies are. The marmoset in question was able to be flexible enough in its behavior to accept an abundance of care when it wasn't biologically necessary; it might have been biding its time until it could make a clean getaway.

The researchers documenting this phenomenon were simply unable to find the marmoset in the capuchin population after an extended time. Perhaps it left on its own; perhaps it died. The bottom line from this and many of the primate examples is that female primates are generally very interested in babies, and when their own bodies are primed for motherhood, they are in the most likely position to either have one or adopt one. In many New World monkey species, females are the philopatric sex, meaning that over time many of the females in a group are likely to be related to one another in some way. Perhaps the abundance of adoptions that happen result from a mixture of chance circumstances against a backdrop of potential relatedness.

Although relatedness is certainly a good reason to adopt an orphaned infant, it's certainly not the only way in which adoption could be adaptive. Adoption occurs in a predictable percentage of many Phocid species including northern and southern elephant seals, gray seals, and Weddell seals. In these animals, heavily pregnant females haul themselves out of the ocean onto beaches or rocky outcrops to give birth, and they do so in large densities. Breeding harems on land can become quite crowded, as females aim for spots in the territories of the most powerful males. They do this to avoid extensive harassment of other males impatiently waiting for their turn at copulation and fatherhood (females mate during their last few days of lactation). Crowded rookeries often mean chaos, confusion, and noise.

Even though mothers generally fast for the three-week period of nursing their pups on land and therefore do not leave their pups unattended for long periods, there are still various disturbances that result in separation of mom and pup or even in pup death, including things like the alpha male moving through the crowd, aggression between females, extreme weather events, or just juvenile pups wandering away from their mothers. Many young females (aged three to five) are unable to successfully rear a pup in these conditions.

In addition to potentially losing their pups, young moms are also the most likely moms to adopt an unrelated one. It's not uncommon for females that have lost a pup to adopt a pup of approximately the same age. There are several hypotheses as to why this kind of adoption may occur, including explanations that are adaptive and non-adaptive. For example, a simple case of mistaken identity could be considered a non-adaptive reason for infant adoption. Mom is simply unable to discern whether the nearby pup is hers or not, and so she errs on the side of providing the intense (yet fast) period of maternal care that many Phocid species require. In general, Phocids have poorly developed mother-to-pup recognition abilities and they are more likely than Otariids such as sea lions to engage in allon-ursing or reciprocal nursing. The three-week lactational phase in elephant seals requires a massive investment from mothers; however, the motherhood period is over once mothers return to the water. Pups are weaned cold turkey, and moms and pups come to exist independently. If an inexperienced mother happens to provide care for the "wrong" baby during one breeding season, it's likely that she gains some valuable experience at providing maternal care anyway. In fact, it could be desirable for inexperienced females to try out their mothering chops on babies that do not biologically belong to them prior to doing the real thing in the following season.

Females may also adopt to maintain a continuance of their physi-ological reproductive cycles or because they are physically and hor-monally primed to start mothering. Lactation and a regular nursing period partially induces the next ovulation, so females that have lost their own pup but adopt an unrelated orphan and provide it with milk will come into estrus, copulate, and give birth in the following season. An early cessation of the lactational process may result in a temporal discord between ovulation, copulation, and pregnancy in mothers that lose pups, so providing lactational support to orphans could be a win for all parties involved. Breeding synchronicity is key in Phocid species, for protection from predators and gregarious sexu-ally mature males, so it makes a lot of biological sense for females to want to maintain their own reproductive cycles in concert with other females despite losing their offspring. This situation represents an adaptive and evolutionarily sound explanation for the existence of unrelated orphan adoption in many seal species.

Although there is not a lot of detailed work on the topic of adoption in Otariid species compared with Phocids, a recent report suggests that adoption does occur in California sea lions to quite a large extent. In fact, 20% of all mom-pup pairs at one study site were biologically unrelated due to the high level of adoption in the population. There are bound to be differences between populations based on any number of demographic, physical, environmental, and social factors at play; however, the existence and persistence of adoption in certain areas is thought to positively contribute to the persistence of populations that are decreasing in size. Pups that are orphaned are unable to exist for long without being "claimed" by another mother, and a rate of 20% of adopted pups in a population, even if this represents an abnormally elevated rate of adoption, means that 20% of the pups in the population were otherwise at risk of losing their lives. On this scale, the fact that these pups were adopted and cared for until weaning translates to stability at the population level. Perhaps if all females take on the role of adoptive mother at some point in their lives, it could compensate for an abnormally high rate of orphaned pup death.

Otariid females generally have highly developed modes of recognition between mothers and pups, so the adoptive mothers are likely to have identified that their orphaned pups are not biologically related to them. However, they nevertheless provide a high level of maternal care. But the larger question is why so many pups are orphaned in the first place. Many are thought to be pups of first-time moms who either abandon or lose them, or pups of mothers that are killed by predators or disease. Adoptive relationships directly mirror biological relationships in this species, indicating that it may be a substantial factor in maintenance of localized California sea lion populations.

Terrestrial mammals like zebras also experience adoption, although it varies depending on the social structure of the species in question. There are three species of zebra: Mountain, Plains, and

Grevy's. The first two species live in long-term stable herds that do not experience either allonursing or adoption. However, Grevy's zebras live in much more fluid social groups known as fission-fusion societies. In this kind of social set-up, different combinations of individuals will meet from a few minutes to several days. When an orphaned nursing infant is part of the fission-fusion grouping, the infant will attempt to obtain milk and care from any of the lactating females in the herd. Because the social groupings are so fluid, the cost to allonurse for short periods of time is not that high for any one mother, so mothers may not prevent it. In addition, mechanisms for mutual recognition are poorly developed in zebras, so one cannot rule out that mothers simply may not recognize whether a foal is biologically theirs or not. Indeed, in many zebra populations (of all three species), mothers will leave their herd to give birth, and return to the herd once a strong bond has been created with their foal. This could be the only way in which a mother is able to recognize which baby is hers.

Infant recognition is also an issue for female eastern gray kangaroos. These animals are gregarious herbivores that live in fission-fusion social societies similar to those of the zebras. Small micro-groupings form at various times with great fluidity. Full adoption is rare in this species, but, when it does occur, it is generally attributed to a case of mistaken identity (given the mother's blunder) and considered to be maladaptive. The rate of adoption in gray kangaroos increases during years when population densities are high, and there are many females with pouch young that are approximately the same size. It's thought that if a lactating mother changes her social grouping and isn't closely followed by her joey, she may lose her infant altogether. The joey would need to find a vacant pouch to literally hop into, and hope that its "new" mother either doesn't notice or doesn't mind; overnight exposure would certainly result in death.

Generally, dependent infant kangaroos remain very close to their mothers during their first several trips out of the pouch. They may take small outings at first, and launch into distress calls if they cannot immediately find their mothers. This all makes good sense for a species that has poorly developed recognition abilities between mom and joey. Most adoptions may occur if a population has experienced a startling or threatening event that requires them to flee the scene quickly. Moms usually perform a "nose check" with their offspring before they hop into her pouch to verify identities. However, if the situation requires immediate action and there is no time for a nose bump between mom and baby, the wrong baby could easily jump into the pouch. Once in the pouch, the joey takes on the scent of the mother, and so it may then be impossible to discern it from non-offspring for the rest of the lactational period. The low level of adoption (3%) in eastern gray kangaroos is related to their population densities—the more kangaroos (and joeys) there are around, the more likely it is for there to be misidentification.

The lack of ability to distinguish one's own offspring is likely to contribute to the occurrence of adoption in polar bears, which are solitary northern carnivores that live in small numbers in the Subarctic and Arctic regions. Females mate between late March and June, and then give birth to one to three cubs in November or December. Females usually give birth while they are in their overwintering dens and emerge with their cubs after a few months. As these are solitary creatures, mothers provide exclusive care to their pups until the offspring reaches independence between two and three years of age.

As with the seals and sea lions, a mother who is physically and psychologically primed to be mothering cannot simply shut that instinct off immediately. But for these solitary creatures with a substantial lactation phase, adoption represents a huge cost. It's thought that the biological readiness plus the lack of fine-tuned recognition abilities both contribute to the occurrence of adoption in polar bears. There are several documented cases of adoption in polar bears that suggest that it is most likely to happen when a mother has lost either all or part of her own litter. A similar situation has also been observed in closely related grizzly bears, where adoptions sometimes occur during the seasonal salmon runs. Grizzlies congregate near fishing sites to feed on the returning salmon, and cubs may either get mixed up, ending in reciprocal adoption, or may be adopted by the "wrong" mother after either being abandoned or lost.

A completely different reason for orphan adoption is observable in populations of wild dogs in Africa. It's quite common for adults to adopt dependent young and gain direct benefits from doing so, regardless of whether the adopted pups are related to them. In some social species, there are direct benefits to be gained from living in a larger group than a smaller one. Individuals in large groups realize a relaxation of predation pressure since many pairs of eyes are constantly on the watch for potential threats. They also realize the benefits of dilution, which means that if the population is attacked by a predator (lions are the most common predator of wild dogs), being in a large group means that it's less likely that any one individual will be the unlucky victim of attack. In addition to the benefits of protection, there are other perks of being a social carnivore in a large group—including increased competitive ability, defense of carcasses from hyenas, hunting success, and reproductive success. In other words, for wild dogs, it makes a more sense to be in a large social group than a small one. So, when orphan pups are encountered, or when a mother with a pup dies and leaves her pup behind, the pack will usually adopt them.

Wild dogs are cooperative breeders, meaning that reproduction is monopolized by a dominant couple and all the other reproductive aged adults in the group are non-productive helpers. As mentioned in the section on cooperative breeding, the helpers are not necessarily related to the breeding couple—which represents a serious biological conundrum when it comes to their own inclusive fitness. The overall benefits to living in a larger group may have somehow selected for the evolution of helping behavior in wild dogs, or at least selected against any kind of intra-group aggression. However it evolved, it represents good news for orphaned pups because it means that they can be rolled into a pack and cared for by adult helpers that weren't caring for their own biological offspring. Parental care in wild dogs extends for approximately a year, although pups are weaned from mom's or allomom's milk at approximately three months. Adding a few extra pups to the group comes at a very low cost to these wild dogs because food is plentiful and the benefits of living in a large group are otherwise substantial.

Overall, the ubiquity of allomothering in the animal kingdom means that moms of diverse species provide care to offspring that are not biologically related. The extent of this kind of care varies both between and within species, and humans are no exception. What we officially term "adoption" could arguably be considered a case of acute allomothering. Regardless of how our own species decides to define it, families, populations, and societies of animals across the planet are better for the unspoken investment of allomothers and adoptive mothers.

13

THE MEANING OF WEANING

There comes a time for all good things to end—and yes, for babies to move toward independence from their mothers. Since lactation is without question the most energetically expensive process for female mammals, the cessation of lactation is a substantial milestone for both mom and offspring.

The question of when to cease lactation is a difficult one to answer for any animal mom, as there is an abundance of diversity among mammals and primates with respect to the individual needs of both mom and offspring. For the most part, it is safe to assume that the benefits of lactation to offspring are the greatest in the first weeks immediately following birth. The costs of producing milk are constant; however, larger, older offspring are naturally able to drink more, so the volume of milk produced by mothers increases to a point of peak lactation before tapering off. At peak lactation, the costs to mom are the highest. It's when baby is consuming the most milk without supplementing their diet with external types of food. After this point, infants are introduced to other food types, and, depending on the species, they are either encouraged to forage for themselves or with help from mom. There are a few physical landmarks that can be loosely associated with the weaning process in primates; however, the main one is the emergence of the first molar. Across species, molar eruption is commonly coincident with the cessation of lactation.

In a perfect world, one would expect babies to continuously increase the level of external foodstuffs in their diet up to a certain point, after which they thank their mothers profusely for the nutrition and investment, and politely stop suckling. However, it's safe to say that this simply doesn't happen. The mother's and the baby's separate needs must be taken into account when considering the cessation of suckling, and unfortunately these two sets are in direct opposition. It's in the mother's best interest to give her offspring the best start at life by providing both nutrition and protection while her newborns are most vulnerable. However, it's also mom's job to consider her own physiological status, as well as her potential for future reproduction. If an offspring is healthy and is physically able to sustain itself, then it's a waste of mom's energy to continue provisioning, plain and simple. This is bad news for said offspring, who would rather have mom at his/her beck and call for as long as possible. If mom is doing all of one's foraging and protecting, life is easier, simpler, and more relaxing. This scenario describes what is known in biological literature as *parent offspring conflict theory* and there are a number of factors to be considered both by species and by individual.

Primate mothers begin to limit access to the nipples at an appropriate stage subsequent to peak lactation; however, the bulk of primates nurse their offspring for at least a year. As their babies near weaning age, mother leaf monkeys will simply reject their infants' attempts to retrieve milk from their nipples, and subsequently ignore the cries and tantrums that come as a result. Much like human infants, baby primates are good at making it known when their needs are not being met, and being deprived of mom's milk is stressful for both nutritional and emotional reasons. However, by ignoring the cries and carrying on, mothers can teach their young infants that their position will not be swayed.

Many mothers make their weaning decisions gradually by introducing new foods and encouraging their infants to both forage for them and manipulate them. The key to weaning for any primate mom is to make sure that their infants feel supported—physically and behaviorally—to insure they are ready for a new level of independence. That said, the outside world can often make this very difficult, if not impossible. In addition to becoming nutritionally independent, juvenile primates have a lot of social developing to do (against a backdrop of fluctuating resources, no less) and they must also consider the ever-present risk of predation. This means that the conflict between mothers and their growing babies is real, and not very clear cut in terms of its resolution.

In many cases, cessation of lactation will occur only once a mother has resumed ovulatory cycling and is pregnant again. For many long-lived mammals that nurse their offspring for years, such as orangutans, elephants, and gorillas, mothers will almost always be pregnant again while still nursing the previous offspring. Weaning conflicts can become particularly intense when this is the case. In rhesus macaques, stress levels in infants (as measured by levels of glucocorticoid stress hormes) are high during the weaning process; infants show both emotional and physical stress from being rejected by their moms. However, the release of glucocorticoids is one way that the body can "help" with the ordeal, as their release results in a series of metabolic changes that ultimately improve one's ability to cope with stressful situations.

For animal moms with shorter periods of lactation such as guinea pigs and meerkats, there can be a distinct and complete cessation of response to pups begging for food. Simple as that. Guinea pigs have a lactation period of approximately three weeks, after which time mom can either continue to feed her pups or simply ignore their calls. At three weeks old, pups are physically able to forage successfully for themselves; however, a milk meal of similar nutrition is easier to obtain. If mom becomes pregnant again right away, she is much more likely to ignore the begging calls of her current litter past the time frame of three weeks. This seems to be a critical stage for guinea pigs, after which all milk provision can be considered "extra." Female guinea pigs make decisions about the end of parental care based on an internal clock which is independent of the pups' age or needs. Moms just "turn off" their mothering behavior at a certain point.

In a similar vein, meerkat pups produce distinct begging calls for food that are recognized as such not just by mom but by other allomoms in the cooperatively breeding group. When pups beg, they are soon after provisioned with food. The specific acoustic components of the pup call, which is generated between forty and sixty days old, is distinct and considered a "food call." At around 100 to 120 days old, juveniles are no longer physically capable of creating the correct "food calls"; hence, they are no longer provisioned by either moms or allomoms. Although it sounds a little like tough love, it's important for providers to not go overboard when it comes to extra provisioning. For most mammals, as soon as the current offspring or litter is weaned, mom can get to work on creating the next one.

There are several ways to consider the plight of a lactating mom when she is determining the course of action she will take in weaning her offspring. First, ecological determinants play a major role. The amount and types of food available is a major factor that is difficult for westernized human moms to wrap our heads around—what with our supermarkets and baby food products and nutritious alternatives for each developmental stage galore. The same is obviously not true for primates in the wild, when weaning an infant at a particular time can either help them to thrive or issue them a death sentence. Seasonally breeding species must take into account the availability of specific foods that are both nutritious and manageable for young juveniles.

For example, female mountain gorillas in Virunga exploit bamboo shoots, a biannually available food source that is high in energy and protein and is easy for infants to manipulate and digest. Bamboo shoots are a good transition food for infants, but they are only available at certain times of the year. This means that offspring of different moms could be at various ages when the shoots come into season. Individual gorilla moms "make the call" about whether to begin weaning their infant during the current bamboo shoot season, or hold on for the next one. There are two predictable "weaning peaks" in the gorilla populations that make use of bamboo shoots, as this is a very important ecological factor for them. Weanlings lack most of their molars, so most the available flora, which requires grinding and chewing, is unsuitable.

On the other hand, tufted capuchins, a small, New World monkey species, have what is called an "extractive" foraging strategy. The foodstuffs that comprise the bulk of the capuchin diet are encased in a hard shell or husk, including snails, bivalves, and nuts, that require both dexterity and strength to extract. The strenuous activity of foraging is particularly difficult for young infants, who need to develop their hand-eye coordination, dexterity, and strength. The required level doesn't happen until infants are approximately eighteen months old, and they remain suboptimal foragers for several years after weaning. This makes the "decision" to wean their offspring even more complicated for capuchin mothers because it represents the beginning of a long period of difficult foraging for her young infants. In fact, many capuchin weanlings lose weight as they begin to exclusively feed themselves, and, in extreme cases, they may even starve. If resources are in such sparse supply that babies are starving, you can bet that the mother's choice to wean her offspring is a matter of her own life and death.

The situation is vastly different for another New World monkey: the squirrel monkey. Juveniles are weaned around five months after birth, soon after they begin to exhibit a very basic repertoire of locomotive skills. Their diet of fruits and small insects is much more conducive to earlier weaning than that of the capuchins. Infants can find and exploit several nutritious food resources using only basic visual and manipulative skills. This illustrates a general pattern of adult diet and weaning. If there are items in the adult diet that are readily available and manipulable by juveniles, this makes a massive difference to their journey of weaning. It's a much easier transition when food alternatives are available. If, however, foodstuffs are either seasonal, difficult to obtain, or otherwise unavailable due to extenuating and unpredictable circumstances, the weaning process is affected.

For example, female chimpanzees that inhabit low-quality foraging areas have a greater difficulty in maintaining the metabolic load of lactation and may not begin to ovulate again until a substantial amount of weight has been gained. This isn't possible if mothers are still providing milk to offspring, so mothers may opt to end weaning to promote their future reproductive potential. Overall, chimpanzees have a very long period of infant dependence; offspring are not fully weaned until they are four to five years old. Solid foods are introduced into the diet during the first year, but most nutrition for the first one to two years of life is received through nursing. After the peak lactation period, mothers increase their rate of rejection of infants looking to suckle, and they dramatically decrease their ventro-ventral (tummy on tummy) contact with infants, which is a common nursing position. Since the period of infant dependence in chimpanzees is so extended, there is generally ample opportunity for mothers to tailor the weaning process of each infant to their specific needs.

Although chimpanzee infants nurse for a long time, they are not the longest suckling babies in the primate world. Orangutan infants suckle for the longest of any primate, generally around six and a half years and, in some cases, up to eight years. Periods of nursing in Bornean and Sumatran orangutans are cyclical and highly reflective of natural variation in local resource abundance. Indonesian swamp forests are notoriously unpredictable environments, and this is reflected in many aspects of the ecology of orangutans. They generally have a low daily energy expenditure to begin with, which is a strategy to avoid starvation during periods of resource scarcity. Infants rely primarily on milk for at least the first year of their lives, and then solid foods are slowly introduced depending on their availability.

By mapping the levels of elemental barium on the teeth of orangutan infants, biologists can ascertain the specific amount of milk provisioning for a given period. Orangutan moms ramp up milk provisioning during periods of resource scarcity to provide nutrition for her offspring, but she will reject attempts to suckle when fruits are abundant. In this way, she keeps her milk supply at the ready while encouraging her infants to become independent. For young orangutan infants, the journey of weaning can be a wavy one, as they can increasingly and decreasingly rely on milk for cyclical periods during their first several years of life.

The situation is not the same for other mammals like ungulates, in which the weaning process begins much earlier than for most primates and varies depending on the ecological practices of hiding versus following. As outlined earlier, maternal ungulates (hoofed mammals) are either "hiders"—who leave their offspring in sheltered hiding spots while they go out to forage—or "followers" whose offspring walk almost directly from birth, and follow their mothers everywhere. As a general rule, hiding species (such as giraffes) wean their offspring earlier than following species (such as wildebeests) because the following infants have higher nutritional requirements than the hiding ones. However, it's important to keep in mind that regardless of the general ecological strategy (i.e. hiding or following), there remain external factors that are likely to alter the weaning process for each infant.

Wild dogs in India provide an interesting example by which to examine mother-offspring conflict and the weaning process. These canines have large populations in urban areas and rely almost entirely on the discarded food waste of humans for their diet. Food competition is intense in these free-ranging dogs, and females have the added burden of bearing a litter of pups every year. In the early weeks after birth, mom exclusively lactates for her pups and she remains very protective of them. However, after the six-week point she will begin to supplement her pups with solids. Around the ten- or eleven-week mark, mom abruptly weans her pups; from this point on, her own pups are competitors for food. It's an interesting conundrum, and the abrupt changeover from nurturer/mentor/food source to competitor must come with a suite of psychological changes for any of the canine moms involved. Wild dog mothers lose a lot of weight when they suckle pups, so it's imperative for them to regain their weight and energy stores soon after weaning. Moms are generally in a very poor state of health toward the end of suckling a litter of pups, but they can regain their body condition by focusing on themselves again once the pups are weaned. In the cutthroat world of street dog life, there isn't time to remain nurturing when one must focus on staying alive.

Maternal survival is a critical issue when it comes to the timing of weaning, but infant survival is critical as well. It doesn't mean much for mom to have gone through the entire processes of pregnancy, birth, and lactation only to wean her offspring too early for them to survive. Sometimes it's more beneficial for mothers to wean their offspring earlier than later. There is a predictable and very stressful situation that many mammalian and primate moms endure which can drastically impact the timing of weaning—and this is infanticide by invading males. As mentioned in a few different sections, many primate groups are polygamous—with one male as the alpha who mates with all the females. Other groups can have multiple males and females, in which a few different males mate with all the females in the group.

Whenever there are alpha males in a primate group, there are also "outside" males that would prefer to be in the position of the alpha. These outsider males attempt to invade the territory of an alpha, thereby inheriting not only the area but the females within it as well. However, when a takeover happens, the new male or males kill all dependent offspring, which serves to bring the group's females back into a reproductive state. This is a generality of reproductive physiology in female mammals—the sucking stimulus of nursing is known to induce postpartum anovulation. A sudden termination of suckling is therefore linked to the resumption of ovulatory cycling and the possibility for reconception. Groups with several males are more difficult for foreign males to infiltrate than groups with only one alpha, which means that there is a general trend of females weaning infants more quickly when the infanticide risk is greater. For example, female mountain gorillas can either be in single-male or multi-male groups, and those in the former wean their infants faster than those in the latter, which could represent a response to the increased threat of infanticide.

Group size in general is another potential factor that impacts weaning age in primates. Phayre's leaf monkey moms that live in large groups tend to wean their offspring later than moms from small groups, possibly reflecting the increased foraging costs incurred from living in a large group with many mouths to feed. A similar trend is seen in house mice, those living in small groups wean their offspring more quickly. Assuredly there are both costs and benefits to living in either small or large groups, and the timing of weaning is but one factor that mothers consider when deciding how to rear their litters. Although weaning happens later in larger groups, there is more opportunity for information sharing and social interaction as well.

Larger groups also carry the potential for a greater level of future competition, especially for the non-dispersing sex. For example, once philopatric, non-dispersing daughters reach sexual maturity they compete with their mothers as sexual partners and for basic resources. Males, on the other hand, represent an "easier" long-term investment because they leave the group upon sexual maturity and continue to represent mom's genetic blueprints without directly competing with her for food. When this is the case, we expect to see primate mothers putting a greater effort into provisioning of their sons before they disperse. Indeed, weaning for sons is often later than for daughters when daughters are philopatric.

Weaning for sons is also later than for daughters when males need to attain a large size to be competitive in later life. In many mammalian and primate societies, males are markedly larger than females; this is called sexual dimorphism. Mothers of sons who need to get big (such as northern elephant seals or Galapagos fur seals) will wean their sons *much* later than their daughters so that they have a greater chance of reaching a large size. This is in addition to the fact that many mothers provide higher-fat milk to their sons than to daughters in the first place.

Weaning is a difficult thing to study and to quantify. Because of the multitude of factors involved that a mother cannot control (resources, predation, environmental stochasticity, number of previous offspring, offspring size, and sex), the time to weaning could easily be different for each one of her infants. In fact, the most important variables in the weaning process involve a multitude of individual characteristics about mom herself.

In mountain gorillas the inter-birth interval of primiparous mothers is 20% higher than for multiparous ones, which isn't surprising given the steep learning curve of first-time moms. Survival of first-borns is lower for primiparous mothers as well; infants born to first-time moms have a 50% higher mortality rate than those born to more experienced ones. Breastfeeding and weaning are just two of the many practices that are difficult when one is doing them for the first time. Primiparous moms generally have a lower quantity and quality of milk than multiparous ones, and this is likely related to both physiological and emotional aspects of each.

The mother's age makes a big difference in the weaning process; younger moms often must consider their own growth in addition to the growth of their newborns, so they may opt to wean earlier and put a greater level of resources into their own growth and development. It's often the case that first-time primate moms are little more than young ladies themselves! On the other hand, multiparous, older moms do not have as much of a requirement to invest in their future reproduction—and so they may choose to wean their offspring later. Many human mothers may reflect that they weaned their "last" baby at a later age because they knew it would be their last chance to experience breastfeeding. It's possible that there could be a similar sentiment in primate mothers; however, most non-human primate mothers continue to produce babies until close to their death. Humans and toothed whales are unique in the animal kingdom in the uncoupling of reproductive and actuarial senescence.

In addition to parity and age, a mother's social rank most certainly plays a role in the weaning time of her offspring. High-ranking mothers have priority access to high-quality foods, and this makes them more likely to attain their pre-baby energy stores more quickly than low-ranking ones. It's been shown for some cooperatively breeding primates that mothers receiving help in the form of allonursing and infant care tend to wean their offspring much more quickly than those who do not (no kidding!). As with most aspects of life, low-ranking mothers have limited access to high-quality foods and are more likely to experience ill-health themselves. For these moms, there is a more difficult trade-off between supplementing their infants and supplementing themselves.

Lastly, a mother's individual personality has a lot to do with when her offspring will be weaned. There is a great deal of diversity in mothering style with respect to how much mothers encourage independence of their offspring. Macaque mothers will play games with their infants as a way to introduce them to independence. They will place their babies on the ground, just a few feet away and then encourage them to make their way over to them. Moms may make familiar gestures or sounds so that their infants begin to understand what is expected of them. Macaque mothers begin these games as early as the infant's first week of life. Meanwhile, female olive baboons, yellow baboons, and Hamadryas baboons all do something similar: put their babies on the ground, sometimes as early as the day that they are born, and take a few steps away before making familiar retrieval vocalizations. Mother chimps take their tiny infant's hands and help them to take their first steps while walking bipedally backward—much in the same way that human mothers help their infants. Perhaps it's not surprising that the females who encourage independence of their infants from an early age tend to have offspring that wean earlier and have more adventurous personalities. However, maternal encouragement is also sensitive to the competence of individual offspring. The fact of the matter is, when it comes to weaning and the encouragement of infant independence, there is no easy way to summarize or conclude that "that's how animals do it." Some babies are simply ready before others, and all mothers must ultimately make this decision based on each infant's ability to thrive on their own.

14

MENOPAUSING MOMS AND GRACIOUS GRANDMAS

A lthough human mothers attempt a multitude of ways to stave off the inevitable with makeup, face cream, or even surgery, aging happens to all animals, primate or otherwise. For those lucky enough to escape the usual suspects like predators, resource shortages, environmental chaos, or disease, the process of aging will most certainly be a contributing factor to their ultimate demise. Senescence, or the natural aging and breakdown of the animal body, is a process that begins from the onset of adolescence. For decades, it was widely assumed by biologists that most animals didn't live long enough to experience old age. It turns out that that's not necessarily true. Challenges like resource acquisition, competition, and predation combined with the physical burden to bear offspring most definitely make it more difficult for animal moms to reach old age—but advances in monitoring technologies have revealed that many of them do.

There are two kinds of senescence—actuarial and reproductive. *Actuarial* senescence refers to the physical processes of aging. Characteristics one would ascribe to an elderly person—wrinkly skin, frail and brittle bones, diminished eyesight—are the work of actuarial senescence. Evolutionarily speaking, the forces of natural selection are not as strong on aging animals, so energies that were once spent on childrearing (for females) or costly sexual signals (for males) are later consumed with keeping an aging body in a relatively healthy condition in the face of normal deterioration. Negative genetic mutations with late-acting effects can accumulate in elderly females, making them even more prone to sickness or injury—which means that aging females may not have either the energy or an adequate level of health to be at her best for offspring she has late in life. This is a substantial problem for almost all mammalian species, because mothers are almost exclusively responsible for childcare. Many mammalian females experience strong selection pressure early in their lives to start reproducing, sometimes even before their own bodies have finished growing. However, there is a myriad of ways in which first-time mothers may struggle, and, in many species, there is a tradeoff between early reproductive success and a faster aging process later.

The other kind of senescence is reproductive, and it pertains to the fact that, in many mammal and primate moms, there is an unmistakable decline in reproductive function with age. Let's start at the beginning. The biological clock starts to tick for many mammalian females as soon as they reach sexual maturity. Given the correct formula of behaviors—culminating in sex with a reproductively mature male—these females become mothers almost immediately. Rodents are a good example of a mammalian group in which females become competent mothers almost as soon as they are physically able to do so. Most maternal behavioral machinery is genetically inherited in rodents, and females are innately competent mothers from young ages.

Female squirrels tend to reproduce from a very young age, as individuals that begin reproducing early generally have a larger overall lifetime breeding success. If there isn't a lot of learning to do, if the bulk of mothering behavior is innate and genetically programmed, then waiting to reproduce isn't a biologically intelligent choice to make. Since these squirrel females have the knowhow and the machinery to make it happen, early primiparity should be the rule. Any kind of waiting period for these animals just increases the incidence of a female dying though some unrelated circumstance prior to realizing any biological fitness.

Indeed, in this kind of scenario, the overall increase in lifetime breeding success usually results in a high rate of senescence at older ages. This is markedly different from the strategy employed by most primate females, who are far from ready to become mothers at the onset of puberty. One of the major reasons for this is that their own bodies haven't finished developing, and to begin sexually reproducing at a young age represents too big of a trade-off between current and future reproduction. In addition, many primate females are not psychologically ready to become mothers at the onset of sexual maturity—and this is especially true for our own species. Imagine if all human females became mothers at age twelve or thirteen! Although their bodies may be physically able to produce an offspring, young human females have much to learn before they can hope to be effective mothers. However, because of motherhood's massive physical demands, one also does not want to leave mothering until too late in one's life. Human females who leave reproduction until their forties are bound to have less energy and feel the impacts of pregnancy in a negative fashion. The bottom line for humans, as well as for most other primates, is that there is a kind of reproductive-age sweet spot that represents a peak rate of fertility. This sweet spot occurs when females are still young enough to have ample energy and resilience for the intense physical demands of motherhood but are old enough to be mature and competent parents.

At this stage in her life, a given female will experience her peak reproductive output, although that doesn't mean that reproduction simply stops after this point. After hitting their reproductive prime, most mammalian females continue to reproduce at a lower rate than they did in their younger years, but they do not stop reproducing until they die. In other words, for most animals there is a relatively equal rate of actuarial senescence and reproductive senescence. Game over for reproduction equals game over for life. However, for a few groups of animals—including humans, short-finned pilot whales, and killer whales—there is a distinct uncoupling of actuarial senescence from reproductive senescence. In other words, the reproductive system ages much more quickly than the rest of the body, leaving both human and whale females completely postreproductive from the time of menopause approximately halfway through their lives. For the most part, a general rule of thumb for mammalian reproduction is that females continue to do it, albeit at lower rates, until nearly the end of their lives.

An important consideration then is the amount of time that a given newborn is completely dependent on its mother for care. It doesn't make any biological sense for a female to give birth to a baby only to keel over the next day and have the baby meet its own death due to a lack of adequate care. That most primate and mammalian offspring require extended periods of care from their mothers as they develop through infancy and toddlerhood means that mothers must be in adequate health to see the entire process through. In other words, if your baby is going to need five to ten years of extended care, it's generally necessary for that baby's mother to have at least ten years of life left to be able to rear that offspring successfully. This critical interval of time is described by the *mother hypothesis*, which posits that there should be a postreproductive interval of survival for mothers so that they can provide adequate care to their last offspring prior to dying of old age. This is especially important in mothers of very long-lived species whose infants may require several years of care to reach a point in their lives where survival without mom is possible.

Asian elephant females provide support for the mother hypothesis. Although they do not experience menopause per se, they do exhibit an extended postreproductive period of at least five years to be able to support their last calf. Elephants have extensive lifespans of up to eighty years, and there is a lengthy period of calf dependence on their mothers for the first several years of their lives. In addition to learning how to survive, forage, and avoid predation, calves require extensive social support from their mothers. Although fecundity is decreased in older elephant matriarchs, a level of reproductive health is generally maintained to see their last calf through to maturity—and this five-year interval is substantially longer than most mammalian females have as a postreproductive lifespan. In this way, it's difficult to conclude whether elephant females undergo a complete reproductive cessation or whether they are waiting out the normal interval that occurs between calves. In addition to reproducing to nearly the end of their lives, matriarch elephant females provide social support to their grown children as well. Male children remain with the elephant group until they reach sexual maturity, at which point they leave and begin a solitary life or join with a group of males. Female elephant children remain with their mother's group for the duration of their lives.

For ungulates such as the chamois and red deer, reproductive senescence in females takes on a markedly different appearance. For these moms, there is a clear and unmistakable decline in the breeding success of older females; however, the bulk of females in the population do not live long enough to experience it. There is considerable variation at the individual level (as there is in many mammals and primates), and so the very oldest females in a population were likely gifted with genes that support their longevity, but there is a clear cost in terms of the adequacy of the mothering that they provide. The gestation period for these ungulates is six months, and calves are normally dependent on their mother's milk for approximately six months following birth. This means that an adult female chamois must trade off her own health and longevity against the probability of successfully fledging another calf—a substantial physiological commitment—for at least a year.

Similar trends are seen in many ungulate and mammalian species. Maternal mountain goats are other ungulates that face the difficult trade-off between investing in offspring *right now* versus waiting for better opportunities in the future. Younger moms tend to have heavier, healthier calves while older moms have smaller calves—a phenomenon which is direct evidence of their lower investment in offspring as they age. The key question is this: when is the health of either the mother or the potential offspring compromised enough for the mother to stop trying to reproduce? When does the risk become too high? There is definitely a trade-off between when a mother should stop reproducing and just concentrate on maintaining her own body for the remainder of her life and when she should try and shoot for just one more successful offspring. This decision isn't an easy one, and the amount of time a mother has left is variable both within and between species. Another complicating factor is something called *terminal investment*, whereby mothers take one last gamble on motherhood.

They provide a substantially elevated level of childcare to their last infant, since it is their last, and that effort may result in their own death. Although it sounds rather dramatic, it's the kind of decision that most animal moms are faced with at some point in their lives.

Studies on terminal investment have mostly concentrated on invertebrate species, where mothers go for broke and can—in some extreme cases—even provide their own dead bodies as fuel for emerging offspring. Things aren't usually as macabre in the vertebrate world, although there is some evidence for increased terminal investment in an array of animals. For example, female Weddell seals are large, long-lived animals that have flexible schedules of reproduction. Mothers provide a huge amount of postnatal energy to their pups during a five- to eight-week lactation period. Moms fast during this time and lose approximately 40% of their postpartum body mass, which is directly passed on to their offspring through milk. As a result, pups usually experience at least a tripling of their body weight during this time. In addition to the significant postbirth investment, young Weddell seal mothers allocate substantial resources to their offspring during gestation. The level of allocation increases with mother age up to a certain threshold, after which it decreases. It's possible that older mothers are simply not able to provide the same level of prenatal provisioning to their gestating pups while maintaining their own bodies in adequate condition to have the pups at all. The reverse is true after older mothers give birth to their pups and start to provide milk: older females provide a greater amount of their own body weight in milk to their pups, which could be an indication of terminal allocation in these animals.

Evidence for terminal allocation and reproductive senescence has also been collected for rhesus monkeys. In that species, older mothers tend to have lower body mass and longer intervals between offspring. In addition, the offspring of older mothers also have lower body mass and survival rates. However, older rhesus mothers spend a substantially greater amount of time with their infants in a ventro-ventral position, as though they are attempting to provide extra love and snuggles to what they realize may be their last offspring. Much of the work done on macaque populations provides strong evidence for both reproductive senescence and terminal investment. Free-living populations of macaques allow researchers to document both the physiological and behavioral changes that take place during the aging process. The fact that there is data to this end at all is quite astonishing, as these kinds of parameters are exceedingly difficult to measure. In fact, this is an important fact to keep in mind—most of the data available for maternal allocation in primates comes from either captive or monitored populations.

It's virtually impossible to obtain reliable measures of maternal investment and physiological cycling in populations of primates that aren't actively managed. Although some primates exhibit sexual swellings associated with ovulation, these aren't always pronounced or easily visible. Not only are maternal investment behaviors vari-able and subtle, but they generally occur beyond human observa-tion, which makes this kind of data fairly difficult to obtain. These data will undoubtedly be influenced by some artifacts of captivity, although they remain interesting in their own right. Since being in captivity generally relaxes both foraging stress and predation pres-sure, it's possible that females in these managed populations may experience increased or prolonged fecundity as a result.

What exactly does a senescing reproductive system look like anyway? What does it mean to say that older females are physi-ologically "less equipped" to bear offspring than young ones? In human females, the aging reproductive system is characterized by a cessation of ovulation and hence menstrual cycling. There is a decrease and eventual cessation of ovarian steroid hormone produc-tion, which means that the body is less able to become pregnant as it ages. Senescent ovaries shrink in overall size, and they become less vascularized, with reduced circulation and gas exchange, which both result in tissue breakdown. The specific hormonal signals that are required to initiate a reproductive event are simply not being fired as often as they once were. This kind of pattern of reproductive slowdown is common across many mammalian and primate species.

Data for the great apes is generally scarce, save for a few detailed studies done on highly monitored, semi-captive populations. But looking at studies can nevertheless be illuminating. Orangutans exhibit age-related depletion of ovarian follicles, gorillas experience longer luteal phases (the period following ovulation), and baboons experience increases in cyclical variation starting at approximately nineteen years of age. Baboons in captivity exhibit a decreased fecundity and increased inter-birth intervals with increasing age, as well as a total cessation of menstrual cycling by age twenty-six. In the wild, it's unlikely for a baboon mother to live long enough to experience full amenorrhea, or cessation of menstruation. Female chimpanzees undergo a phase of reduced fertility akin to the human female's perimenopause, and they also undergo a complete cessation of reproduction if they are lucky enough to reach an old enough age to experience it. In many primate and mammalian species, females may experience a short menopause where they are no longer reproductively active at all; however, for the most part this is either short or nonexistent due to extrinsic pressures on mortality such as resource availability, environmental unpredictability, or predation. Human females are generally free from this kind of threat to their survival, so the experience of complete menopause is a normal event.

Perimenopause is a normal occurrence in both chimpanzees and humans, and it is marked by a lengthy period of irregularity in menstrual cycling that ranges from two to eight years. Endocrine changes associated with perimenopause in chimpanzees include a gradual increase in follicle stimulating hormone (FSH) and leutenizing hormone (LH) concentrations. In fact, by the time human females hit menopause, the levels of LH in their ovaries have typically more than doubled. Following these physiological changes associated with perimenopause, human females undergo a complete cessation of reproductive cycling and they remain otherwise healthy and functioning for an extended time—up to many decades. Surely this isn't merely an artifact of human females being free from the pressures of environmental stochasticity or predation? The short answer is no; it's not an artifact, and there are well-founded theories for the existence of postreproductive females in both humans and toothed whales. The bottom line for chimpanzees, however, is that females experience reproductive senescence in the form of decreasing menstrual cycle frequency. When females are between ages fifteen and thirty, their ovulatory cycles are approximately thirty-five days in length and they have ten or eleven cycles per year. However, variation in cycle length begins to increase at age thirty such that after age thirty-five, cycles are forty-five to fifty days in length and females experience seven to nine of them per year. Associated with the overall increases in cycle length are surges in FSH and LH, which again is akin to the kind of processes experienced by human females that signal a decrease in fertility.

In chimpanzee females that are over forty years old, there is a dramatic cessation of menstrual periods—termed reproductive amenorrhea—and it's the last step before a full shutdown of reproductive function like the one seen in our own species. Full menopause is generally defined as an amenorrhea of over twelve months, and the overwhelming majority of chimpanzee females do not live long enough to qualify. Researches who have studied this phenomenon have done so using captive chimpanzee populations where individuals likely have extended lifespans due to the relaxation of resource availability and predation pressure. In wild populations, females are likely to experience the effects of perimenopause or reproductive senescence to a certain extent, and most females will continue to reproduce (albeit to a lesser extent) during this time.

Macaque females cease menstruating at approximately twenty-five years of age, and many other primate females (such as gorillas, bonobos, and baboons) experience either increased variability of their ovarian cycles or an increased interval of time between them. Although this is a notoriously difficult process to examine in detail, we can conclude with a degree of certainty that many primate females experience a reproductive slowdown toward the end of their lifespans—that is, if they have been lucky enough to live to old age in the first place.

The obvious question then is why there is such a discord between the age of reproductive senescence in human females and their overall lifespan. After all, human females have a lot of living left to do once their reproductive cycles cease to exist. Why would evolution condemn them to a place of biological irrelevance? The truth is, it doesn't. Though part of the increase in human lifespan can be explained through the development of modern medicine and the relatively benign living conditions of the contemporary world, women in traditional societies and hunter-gatherer tribes also experienced menopause at roughly the half-way point in their lives. Living beyond the age of sixty in modern hunter-gatherer populations is much more common than was once assumed, and daughters of postreproductive mothers begin to have children both earlier and more often than those without.

Human females—both today and throughout history—are nearly unique in the entire animal kingdom in that they can continue to increase their own biological fitness by contributing to the reproductive success of their daughters. The *grandmother hypothesis* was coined to describe the possibility that human females realize a good deal of biological success by putting their efforts into the collective success of their daughters' offspring. As stated previously, most primate females have an increasing fecundity and reproductive success up to a certain age. After that point, their own reproductive success comes at a greater cost since the maternal body is past its prime and the massive physical and emotional commitment that it takes to have a baby remains the same. It's an enormous investment, and it only becomes more massive for an aging body. At some point, the cost for a given aging female to become pregnant and undergo the entire maternal process simply becomes too high for her to continue to do it and maintain her own physical health. However, all is not lost. A female can continue to achieve increased biological fitness by conferring her help on the offspring of her adult daughters. The environmental circumstances experienced by our ancestors in the African savannahs contributed to this solution in a totally different, yet complementary way.

As ancestral humans were expanding across the African plains during the Plio-Pleistocene era, they experienced a massive change in the resource base that is thought to have contributed to the evolution of grandmotherly care. The environmental conditions were increasingly arid, which directly impacted the ability of young ancestral humans to provide food for themselves. The resources underwent a change from soft, easy-to-handle vegetation to roots, tubers, and underground sources of nutrition that were impossible for tiny weak hands to obtain. Mothers in this era were faced with an important decision: follow the retreating food sources that their children could obtain for themselves, or stay in the arid savannah areas and ramp up their own provisioning efforts instead. A mother would realize a distinct advantage if her own mother was available to help provision her offspring. Instead of mom having to decrease her direct child-rearing efforts, she could leave the care of older infants to grandma.

The situation represents a win for the mom—who gets extra help in the form of food provisioning, social support, and other offspring care—but it's also a win for grandma, who can continue to increase her biological fitness without having to have another offspring directly (which, as already discussed, becomes harder and harder for an aging body). Vigorous grandmothers who provide a high level of care are especially likely to have a higher number of grand-offspring, and therefore have a greater level of representation in future generations. As "fit and active" grandma genes become more and more prevalent in populations, more females would offer grandmotherly help instead of continuing to birth their own babies, eventually resulting in what we now know as the evolution of reproductive senescence or menopause. The *grandmother hypothesis* is so named because it only works along maternal lines of inheritance. Any mother is sure to a good degree of certainty that the children that emerge from her vagina belong to her. Similarly, a grandmother can be certain that she is the grandmother of any offspring to emerge from her daughter or daughters. The same level of relatedness-certainty is simply not possible for either fathers or grandfathers, which is primarily why most mammalian and primate fathers don't stick around to help in the first place. If a father plays no role in the bringing up of his own offspring, he certainly won't be around to help in tending to his grand-offspring either. Continuing along the lines of this argument, there could be evolutionary differences in the amount of provisioning that grandmothers give to different grand-offspring based on their level of relatedness.

The *X-linked grandmother hypothesis* is an extension of the *grandmother hypothesis* that is concerned with the level of relatedness between grand-offspring and grandmothers in their X chromosomes. Females always contribute an X-chromosome to their offspring, regardless of whether that offspring becomes a boy or girl. If her male partner contributes an X-chromosome, the offspring will be a girl; if he contributes a Y-chromosome, the offspring will be a boy. Thus, maternal grandmothers are 25% related in their

X-chromosomes to both their male and female grandchildren because they contribute one X-chromosome to their daughter (who also receives one from her father) and then one of these X-chromosomes is passed to both her male and female offspring.

For paternal grandmothers, the story is a little different. Male children of a given mother will have inherited their X-chromosomes from mom. Daughters of this male will have this same X-chromosome since that's the only one that the male has. This means that paternal grandmothers are 50% related to their granddaughters when it comes to the X-chromosome. Paternal grandsons inherit the Y-chromosome from their father (not his X-chromosome), meaning that a paternal grandmother is 0% related to her grandsons when it comes to the X-chromosome.

This is a complicated web of relatedness between grandmothers and all their assorted kinds of grandchildren. How might such differences in relatedness matter in the real world? Is it really possible for grandmothers to play favorites based on their relatedness to specific individual grandchildren? At least a few data sets show that this appears to be the case. A metadata set comprising population statistics for seven traditional human populations varying in geographical (Japan, Germany, England, Ethiopia, The Gambia, Malawi, and Canada) and temporal space (seventeenth to twentieth centuries) demonstrated that males realized greater survival in the presence of maternal grandmothers, while girls realized better rates of survival in the presence of paternal grandmothers. It's possible that this kind of recognition could be based in pheromonal cues or on physical likeness between grandmothers and grandchildren. In any case, it seems unlikely in modern human societies that grandmothers make specific decisions about grandmotherly care based solely on genetic relatedness. For the most part, human grandmothers are providing important and biologically relevant care to their grand-offspring without any kind of favoritism.

The only other group of animals that experiences menopause to the extent that humans do is the toothed whale. Short-finned pilot whale and ecotype resident killer whale females can live to the age of ninety; however, they lose their ability to reproduce at approximately the mid-point of their lives in much the same way that humans do. For killer whale females, 95% of their own lifetime fecundity has been accomplished by the time they are thirty-nine years old. The way that these orca societies are formed is unique— and quite distinct from most mammalian populations.

Killer whales live in matriarchal groups, and both male and female offspring remain with the group until their deaths. There are different matriarchal groups that maintain frequent contact with one another; these are called pods. This ecological setup is quite unusual in comparison to most other mammals; in most cases, once offspring reach sexual maturity, either the males or the females split off from the natal group. When one sex disperses like this, it helps the entire population to avoid inbreeding as well as to minimize reproductive competition between members of the same sex. The *reproductive conflict hypothesis* suggests that the cost of competition between mothers and their adult children (of the non-dispersing sex) is an important consideration when mothers and daughters remain in the same social unit and potentially compete for mating opportunities. However, this isn't the case for orcas, even though all offspring of both sexes remain in one group.

Killer whale groups exhibit something called *exogamy* when it comes to mating. Mature males and females temporarily leave to find a mating partner; they mate and then both subsequently return to their natal groups. This is where the killer whale scenario gets very interesting. Females that come back to the group are pregnant, whereas males that come back to the group have impregnated a female in a different pod or clan. Overall what this means is that adult females bring more "work" to the entire group (especially for grandma, who plays a leadership role) in the form of that calf that will remain with the group for the duration of its life and require a great deal of care. Everyone's workload increases as group size increases, including duties such as food gathering, defense, and social function. The new calves do represent an increase in the biological fitness for grandmothers who are a quarter related to their grand-offspring, but there are strings attached there in terms of increased labor. Adult males also do their part to increase a grandmother's collective fitness; however, she doesn't have to do any additional work for it because her son's calves will be raised in another pod with help by a different grandma. Bottom line translation: adult sons are the best bet for grandmothers to increase their biological fitness because grandmothers don't need to do any additional work for it. This is reflected in the amount of help afforded to a matriarch's adult sons versus her adult daughters.

Adult male killer whales are the world's biggest mama's boys. Literally. Matriarchs strongly protect their adult sons, keep them free from agonistic interactions with other groups or individuals, and make sure that they get enough food. Although they also provide help to their adult daughters as well, the vast majority of killer whale effort goes to their sons. Maternal food-sharing with their daughters declines when the daughters reach sexual maturity and begin to have their own calves, but the provisioning of sons goes on to their old age. In fact, an adult orca male over the age of thirty experiences a 13.9-fold increase in his mortality risk in the year following his mother's death. This is a massive number and indicates the clear importance of the provisioning of adult orca males by their moms. Females experience an increased mortality risk as well (a 5.4-fold increase), so although there is disparity between the amount of help allocated to sons and daughters, there is no denying that killer whale matriarchs provide help to their adult children.

The kind of help provided by killer whale grandmothers can take many forms, from group defense and communication to finding and obtaining enough food. There is accumulating evidence that the cultural knowledge passed on by elders is also a very important means by which younger group members learn how and where to forage. The ecological knowledge that is shared by the elder matriarchs about food sources, especially during times of resource patchiness or environmental stochasticity, is critically important for the success of the entire group. The main food resource for resident orca populations is salmon, and female elders easily have a wider body of knowledge about how and where to find them when times are difficult.

Another fact that promotes grandma's ability to help is that orcas are a food-sharing species, a rarity in the animal kingdom. Individual whales could easily consume whole salmon for themselves, but they do not. Fish are broken up and shared between individuals at all times, and so salmon represents both a food source and a means by which to solidify group cohesion.

Does the provision of cultural knowledge help explain why grandmothering evolved in these whale groups? There is a great deal of evidence to suggest that a female can continue to bolster her own biological fitness in indirect ways through leading her group and provisioning her adult children. Such assistance is especially helpful and important during periods of resource shortage and stress. Postreproductive females that act as key repositories of environmental and social information can be seen as a kind of buffer for specific pods against unpredictable environmental hardships. The opportunities for a given female to do this increase as she ages, because as older pod members die off, they will be replaced by her own (non-dispersing) adult sons and daughters.

Upon first consideration, one might conclude that human and toothed whale females become somewhat biologically irrelevant for the latter half of their lives. Aging is an inevitable part of life; however, reproductive senescence is a unique biological phenomenon that plays to an aging females' *strengths* rather than her weaknesses. Indirect help provided by grandmothers is an indispensable part of motherhood for many humans (myself included). The bottom line is that grandmothers offer help not only with day to day functioning (without compromising their physical health with the colossal tasks of gestation and childbirth), but also with the wisdom and knowledge they have to offer. Grandmothering can be considered evolution's way of maximizing the best possible qualities of aging females by taking the hardest parts out of the equation.

15

MOTHERS IN MOURNING

For mothers, there is nothing more painful than losing a child. The unfortunate truth in the animal kingdom is that it happens all the time in a variety of circumstances. We can call this type of death the adverse result of unpreventable circumstances; there is really nothing that a mother can do to save a sick child from a virulent pathogenic infection, for example. This is just the reality of any mother's life. Though it is rare, some babies are born with a kind of malformation or handicap that prevents them from surviving in the long term. Other times, babies are subjected to predation by larger and more dangerous species. Mothers do their best to remain vigilant on behalf of their babies and infants, but it's not always possible to be 100% effective. After all, many large predators (such as lions and hyenas) are evolutionarily predisposed to prey on younger, weaker animals—and moms often witness their own offspring being killed and eaten.

Another major source of death for infants is sexually selected infanticide by alpha males. Many mammals and primates have polygynous mating systems in which there is one male with a harem of females. The alpha male is the father of all offspring of the females in his harem; he is unquestionably the leader and mates with all females in turn. However, if the alpha male is challenged and defeated by a new alpha, this new leader will kill all the infants in the group because they are not his own offspring. Not only does this help to minimize the effort he will exert toward babies that are not his genetic sire, but it also serves to bring the females back into a physiologically reproductive state more quickly so he can begin building a tribe of his own. In this case, the insult to females is twofold—not only do they have to stand by and observe as the new alpha male kills their offspring, but they are then obliged to copulate with him and to bear his young.

It is difficult for us to know what animals think of death. Does any animal truly understand the meaning of death? What aspects of death can be appreciated and understood by animals other than humans? After all, there are many highly intelligent creatures out there with cranial capacities that lend themselves toward sophisticated thought processes and cognition. Although it's impossible to definitively ascribe cognitive traits to animals, there are certainly behaviors and actions that lend biologists to believe that the behavioral reactions of mothers to the deaths of offspring show that they understand, as we do, the four characteristics of death: inevitability, irreversibility, non-functionality, and causality.

Since there is no real way to test whether or how moms mourn their babies, the research on this subject is composed of a rather small set of detailed observations of what moms do in the wake of losing a child. Not only is this not the primary subject for most people that are out there working with animals, but the grieving and mourning processes exhibited by animal moms are usually accomplished in privacy and to catch one in the act (so to speak) is difficult and rare. For these reasons, what we know about how mothers react to their infants' death is based on a small number of first-hand observations. There are a few animal groups that are the most often observed when it comes to the practice of mourning. These are mostly social mammalian species.

Giraffe moms form strong bonds with their calves and remain in close contact with them for the first twelve to sixteen months of their lives, only leaving them briefly—and only while they are well-hidden to protect them—to procure food and water for them. Mothers and daughters will often remain closely associated for many years. Giraffe moms tend to remain close to their deceased infants for an extended period postmortem. However, in the various African landscapes that they inhabit, dead bodies do not sit around for long before being processed by scavenging mammals and birds. In one case where a female's calf was fatally attacked by lions, the mother remained close to the kill site, approximately 650 feet away. She watched the lions consuming her calf while she sniffed the ground and remained on site for four days before moving away.

In another case, a calf that had apparently died of natural causes was closely guarded by its mom. On the morning that it died, the mom was accompanied by sixteen other females, all of whom stood watch around the body. Some of the females nudged the carcass or sniffed it. By the afternoon of the calf's death, there were twenty-three adult females and four juveniles accompanying the mourning mom at the body. By the evening, the mother and fourteen females remained clustered there. Although the number of accompanying females eventually decreased to zero, the mom remained steadfastly with her calf until the fourth day, when the carcass had been completely consumed by scavengers.

At the Soysambu Conservancy in Kenya, a series of observations was made of a female and her calf that had been born with a physical malformation of one of its hind legs. The calf remained alive for one month, during which time the mother was never further than twenty meters away from it. The calf was unable to walk properly and required direct provisioning from its mom. After the calf died, the mother and baby were surrounded by seventeen female giraffes who appeared agitated and extremely vigilant. The mother repeatedly inspected the calf, smelled it, and nudged it. Although the bulk of female visitors moved on within two days, the mother remained with her calf for an additional two days, during which time she continued to inspect the corpse.

The previous examples show that mother giraffes have a clear idea about death, and the fact that they remain steadfastly at the side of their dying or dead calves is a testament to the strong emotional bonds they have with them. The gatherings of conspecifics (members of the same species) that can potentially be observed as supportive are attended only by adult females. Males always keep their distance from the deceased calves—not associating with them or with the gathered females gathered with the calf's mother. During these mourning phases, none of the giraffe moms were observed eating, indicating that a period of fasting accompanies the death of a calf. Perhaps that kind of thing goes without saying—what mother would possibly feel hungry at a time like this? Adult female giraffes generally consume approximately seventy-five pounds of food per day, so not eating for four days represents quite a massive amount of lost energy.

It is commonly accepted that elephants are extremely intelligent and curious animals, capable of complex social interactions. Recorded events have clearly demonstrated that elephants are intensely curious of other elephant corpses at varying states of decay. Mourning has been observed in response to conspecific deaths, and it most often involves investigation of the corpse and a display of agitation and trumpeting. The specifics of a mother reacting to a dead calf are not well recorded, although, as with the giraffes, females will stand with her and remain with the calf for extended periods. Adult females remain in an evident state of distress around dead calves.

In a somewhat fortuitous turn of events, a research group in the Central African Republic observed that an infant forest elephant had become extremely weak. The calf remained close to its mother, who rejected the calf's feeding attempts. The mother instead fed an older calf, who also remained close at her side. The small calf starved to death, and the mother and her remaining calf stood over her for an extended period. The mother sniffed and investigated the body. After the first day, the mother and remaining calf left the area and were not seen again. A total of ninety other individuals came to investigate the body, which was located next to a well-utilized trail and was therefore likely to be spotted. Many of these individuals, both male and female, returned to the body a number of times and engaged with it in a number of ways. Several of them performed investigative-type behaviors such as sniffing and touching it with their trunks. A few adults attempted to lift the small body with their trunks or legs, and a few appeared extremely agitated and even made attempts to defend the body against other visitors. The observing researchers made the important point that the individual personalities of each visiting elephant made each encounter unique.

In much the same way as it would in our own species, each elephant reacted to seeing the body of a young calf in its own individual way. But of course this process must have affected the mother the most. The unforgiving and harsh truth of the matter is that sometimes environmental conditions dictate that a mother simply doesn't have enough resources to feed more than one off-spring. Many moms exhibit a bet-hedging strategy that means they can have a single offspring when times are bad or the chance at two offspring when times are good. However, in making this sort of choice, a very difficult condition can present itself, as when mom loses that bet and one of her infants starves to death. Since mom has already put more effort into an older offspring, it makes more sense, biologically speaking, for her to nurture the older one and let the younger one die.

It's all well and good to read about that kind of biological strategy in a textbook, but it's hard to imagine what that must be like for mammalian moms that are living it. Gestation in elephants is nearly two years, and infants remain close to their mothers for at least another two. This means that the mom in question had a history with the dying offspring of over four years—more than enough time for important emotional bonds to form. The thought of making that impossible choice and then following through with it by starving your own child is hard to fathom. However, these examples show that animals with complex emotions and social bonds must make impossibly difficult decisions that are based in biology.

Toothed whales called *Odontocetes* are mammals that form close social bonds, most often along matriarchal lines, and it is therefore not surprising that an infant's death can be traumatic for a mother. Many observations of different species show similar behaviors of mothers whose calves have died. Most frequently, these include carrying the calf on their backs. She will drape her calf over the anterior part of her back, close to the dorsal fin, and she'll be extremely careful to remain close to the surface of the water so that the corpse doesn't fall off. If the calf slips off, she will go back and pick it up again and carry on the ritual. Such behavior has been exhibited in Risso's dolphins, bottlenose dolphins, beluga whales, and humpback dolphins. Other cetaceans such as killer whales, sperm whales, and short-finned pilot whales carry the carcasses of dead calves in their mouths.

In many cases, the deceased calves are pushed and nudged along by assisting females, despite in some cases being in an advanced state of decay. In observations of an Australian humpback dolphin (detailed in Reggente et al. 2016), the mother always remained within one body length of her deceased calf and repeatedly attempted to hold it at the surface of the water. In one case in which a Risso's dolphin in the Red Sea was observed carrying a calf, humans managed to rope the calf and drag it to shore. The adult female remained with the calf until it was in very shallow water, and she then remained in the shallows for quite some time afterward. In another case of human interference, a bottlenose dolphin mother was startled away from the calf she was carrying by the observing research team. The scientists remained nearby but still for two hours, and the mother returned to the calf and resumed her postmortem corpse-carrying.

These examples show that the mourning process in cetacean moms can be quite extensive. It is difficult and rare to observe due to the aquatic environment, but it seems that since the degree of decay of the dead calves varies to such a great extent, some mothers carry their deceased offspring for substantial periods of time. Newborn dolphins predominantly swim with their mothers in what is called the *echelon* position, in which the calf is flanking their mother's dorsal (back) region. This is advantageous because it keeps newborns in close contact with their moms while at the same time reducing their transport costs. They are essentially carried along by a pressure wave that is created by the mother's body, and this allows both mom and calf to continue moving with the speed of their pod. The fact that so many individuals have been observed carrying dead infants in a modified version of the echelon position is not surprising;

it is likely a natural reaction to providing extra help to the baby anyway. The behavior is clear, but the internalization of death is not. Are these mothers aware of their calf's predicament? Where does the conscious realization of death come in for a cetacean mom? Researchers seem no closer to a final answer.

Primates are the most well-observed group of animals when it comes to the topic of a mother's reaction to offspring death. Cases of continued postmortem maternal care have been observed across several species of monkeys and apes, and many show similar patterns. Essentially, when an infant is newly deceased, mothers go through a phase of continued care, as though the infant is perfectly fine. Over time, the lack of response of the body, in addition to the state of decomposition, provide unmistakable signals to the mother that this baby no longer requires care. However, the variation in postmortem care that mothers show indicates that the behaviors may not be related to mistaken interpretation by moms. They may instead be related to a lengthy and ritualistic mourning process.

A typical two-phase reaction is demonstrated by many primate moms in response to deceased offspring. Upon the death, mothers experience a period of intense agitation, movement, and vocalization. It's a kind of protest behavior that is likely an expression of anguish, shock, and disbelief. During this phase, the mother is engaged in intense interaction with her baby. No other conspecifics are tolerated nearby, and they are certainly not welcomed near the baby. After this initial period of intensity, many moms then go through a stage of silence and immobility. They often sit with a hunched posture and exhibit marked reduction in any kind of environmental interest. This seems to mimic depression-like symptoms, and this stage is arguably a similar kind of reaction to how human mothers would act following the death of a child. The effects of the depressive state can be somewhat mitigated by the presence of a social group, and in the primate world this is most often other adult females. In fact, adult females (not adult males) may often take a turn carrying the deceased infant of a friend or family member. There are many aspects of motherhood both joyous and mournful that can only be made better by the presence of other females who understand.

After the initial phases of intense grief and depression, primate mothers continue to remain in close contact with their deceased infants. As time progresses, they gradually transition from extreme attachment to a more relaxed state of attachment in which other individuals are permitted to view or investigate the body. The behavior of primate mothers in response to their dead infants is varied between species; however, it most certainly shows two of the four factors that are involved in our human understanding of death—that of its irreversibility and its non-functionality. Since the changes to their infant's behavior involve both factors, there is without question some understanding that they will no longer be what they once were. From this, it can be concluded that at least a partial understanding of death is experienced by bereaved mothers, and this contributes to their prolonged rituals with their infants.

Chimpanzee mothers in Bossou, Guinea, also carry their deceased infants for extended periods of time. When a two-and-a-half-year-old infant died from a respiratory illness in 2010, the mother carried its corpse (which became mummified as it aged) for at least twenty-seven days. During this time, she provided extensive care to the small body, grooming it often and sharing her day and night nests with it. The mother also showed marked signs of distress during the few times she was separated from her baby's body. The continuous care of a long-dead infant is a sure testament to the strong emotional bond between mother and child. Several other instances of similar nature have been observed at Bossou, Guinea. During the dry season in 2003, the chimp population experienced a respiratory illness outbreak that left many members dead. Among the deceased were two infants—one just one-and-a-half years of age and the other a hair over two-and-a-half years old. In these cases, both mothers carried their dead infants for sixty-eight and nineteen days respectively, and the corpses underwent complete mummification.

The fact that mummification sustains the infant's corpse for lengthy periods of time makes it possible for their mothers to carry them for so long. It's been hypothesized that this kind of maternal mourning ritual is only possible in environments that lend themselves to the mummification process. Bodies remain encased in leathery skin, more or less intact. In all three cases of extended postmortem maternal care observed at Boussou, the mothers were rarely without their infants, and they chased away flies that circled the corpses. This, in addition to the continuous grooming and the arid environmental conditions, facilitates mummification.

Other group members of all ages and both sexes also exhibit interest in deceased infants. While in their mothers' care, other chimpanzees investigated the bodies, touched and smelled them, and were sometimes even permitted to handle them. In a few rare instances, curious juveniles and infants were allowed to carry the corpse short distances away from its mother, and play with it. It's quite surprising to me that the corpses elicit such interest, and not aversion, from other group members. Though the carcasses were putrid and smelly, there was no aggression toward them. Mothers were never far from their babies, and they made sure to retrieve them to let others take their turn in the ritual.

Mother chimpanzees may also use tools in efforts to groom their dead infants' corpses or keep pests away. This is evident from observations of mothers using branches to shoo flies, and also from a curious incident in which a mother used a piece of grass to clean her dead son's teeth. She worked with intention to clean his teeth while a daughter of hers stayed close by. In this instance, the rest of the chimpanzee tribe had been lured away from the area by rangers using high-quality food. What this indicates then, is that mom (and the daughter that stayed by her side) made a conscious choice to remain with the deceased and groom him rather than partake in food rewards. The treatment of corpses by chimpanzees is meaningful and unique. This should not be surprising to us, since we share most of the same genetic code.

Gelada baboon mothers have been observed to carry their deceased infants for extremely varied amounts of time—less than an hour to over forty-eight days. The mother who carried her infant for the longest period continued to do so long after most of the flesh had rotted away from its skull. As with the chimpanzee mothers, geladas continue the grooming process with their infants, and other group members show marked interest despite the smell of the decaying corpses. When gelada mothers carry their babies for extremely long periods of time, they no longer use the same carrying methods that were used when the babies were alive. Instead of carrying them dorsally or ventrally, long deceased babies are carried with one hand or in the mouth. This is quite important, as it signifies that there must be a cognitive understanding that the babies are not alive, yet the mothers continue to keep them close. In the case of the mother who kept her deceased infant for forty-eight days, copulation had been resumed approximately two weeks before she abandoned the corpse. She repeatedly copulated while continuing to clutch the dead baby in her hand. Although it's impossible to understand the cognitive state of this female, there is no one way that any individual mother may react to the death of her child.

Indeed, multiple cases of primate mothers carrying and providing care to deceased infants have been observed. From chimpanzees and geladas to spider, squirrel, Japanese, and snub-nosed monkeys, mothers of diverse species have been known to carry and care for deceased young. There have been several hypotheses proposed as to the existence of postmortem maternal care. First is the *postparturient condition hypothesis* which states that the circulating hormones in mom's postbirth body influence how she reacts to a deceased infant. Hormones clearly play a large role in the formation of the mother-offspring bond to begin with, and the initiation and maintenance of lactation, infant carrying, and other associated behaviors. What's not entirely clear is the role that hormones play in a mother's body once the baby has died. Is there an abrupt hormonal change or does it taper off slowly? How do each of these scenarios look from the point of view of a mother that continues to provide care to a deceased infant? Since there is a massive amount of variation in the length of time that mothers show care to deceased infants, there cannot be an easy answer when it comes to the role that hormones play. In some chimpanzees, mothers abandon the corpses of their infants well before their hormonal states return to pre-birth levels; however, in other cases mothers hold on to their offspring long after menstrual cycles resume. There is clearly much to be said for individual variation in the mourning process.

Second, the *slow-decomposition hypothesis* suggests that in environmental conditions conducive to mummification, mothers may continue to provide care to infants because the onset of physical conditions associated with death may be delayed. Eventually, the signals of body decomposition become abundantly clear, but it's possible that there could be an early phase of confusion with respect to the status of the infant. Associated with this hypothesis is the *unawareness of death hypothesis* which essentially argues that mothers and other group members may not be able to *instantly* distinguish dead infants from live ones. If it's not inherently obvious one way or another, a mother who continues to provide care will be rewarded with her infant's life if the infant was in a period of infirm or stillness but not dead. The latter hypothesis could explain continued maternal care for infants in the few days postmortem; however, the intelligence levels of primates far exceeds that which would be required to determine whether a decaying baby is still alive. The ubiquity of the behavior across species, and the complex emotional and cognitive capacities of primates, indicates a pattern of infant mourning and death ritualization.

The continued carrying and care of deceased infants is a clear and well-established pattern of behavior that occurs across many primate species. What is more difficult for us to understand are the emotional correlates of this behavior. When and how does a mother give up and finally abandon the body? Are there phases of grieving that must be addressed before such a thing can be done? Perhaps one of the largest and most difficult aspects of this situation to address is that different mothers are likely to handle the deaths of their infants in very different ways, just as human mothers who each mourn in their own way.

Another factor to consider is how the baby died. Was it a lengthy illness or malaise that may result in an ultimate feeling of relief for a worried mom? Is this mom's first child (in which case she may be slower to realize that her deceased offspring is not behaving like a "normal" baby), or has she had other babies before (and realizes immediately that this one is acting abnormally)? Was it a sudden death by a well-disguised predator, and if so, is there a feeling of guilt involved with a mother's sadness? How about accidental death? In the case of observed infant deaths in Barbary macaques where a four-month-old female was hit by a car but didn't immediately die, her mother immediately retrieved her injured infant from the road and took her up a tree. The mother began extensively grooming the young female, but she succumbed to her injuries soon after. In addition to carrying and grooming her infant, the mother was also observed to self-suckle following her infant's death. This has been repeatedly observed in Barbary macaques and may be a means of stress relief for bereaved mothers.

In addition to the myriad factors that could influence the maternal bereavement process, the physiological nature of the loss should be acknowledged as well. If an infant is still very young at the time of death, a mother will also experience hormonal changes associated with the cessation of lactation (i.e. the postparturient condition hypothesis mentioned earlier). How do these hormonal changes factor into the grieving process and do they affect the length of time that postmortem maternal care is applied?

Another aspect of death and the impact of motherhood is how a mother's offspring react to *her* death. In nature, most animals that are newborns or infants at the time of their mother's death are almost always destined to meet the same fate unless they live in a cooperatively or communally breeding species where they will be cared for by another mother figure immediately. However, we don't know too much about how older infants, juveniles, and adolescents fare after losing their mother. Chimpanzee mothers provide a good deal of emotional support to their sexually mature sons, even though the support isn't necessarily direct (i.e. lactation, transport).

In chimpanzee populations, males are the philopatric sex, meaning that they remain in their natal group for the entirety of their lives. When they reach sexual maturity, females disperse from their natal populations and find a new group in which to mate and give birth. Once an immigrated female has given birth, she tends to remain in that group for the duration of her life. This means that mothers and their male offspring cohabit for the bulk of their lives. Although their relationships go through changes as the sons get older, mothers can continue to provide social support, food, and grooming to their adult sons. When an adult male has his mother around, he is more likely to obtain an alpha male position in the group.

Mothers most certainly provide social support to their sons, and this ultimately leads to an increase of both her son's and her own reproductive success. When male chimps lose their mothers before reaching five-and-a-half years of age, their chances of losing their own lives as well are great, no longer receiving milk or sharing a night nest with their mothers. This argument is not the same for female chimpanzees, who disperse from their mother's group at sexual maturity and are therefore completely independent. Male chimps remain dependent on their mothers for support well into their adolescence and even into adulthood, so, if they lose their mother along the way, their lives change measurably. Younger adolescents and juveniles will demonstrate a reduced life expectancy, and older males are not as likely to gain the same social and reproductive standing as they might have otherwise.

There is a similar pattern in killer whales, where postreproductive mothers provide additional support to their adult sons (above and beyond the support that they provide to their adult daughters) for their entire lives. An adult male killer whale has a drastically increased chance of dying in the year following his own mother's death. What does this ultimately mean? It shows the intense and unique power of the mother-offspring bond. Regardless of whether she provides direct or indirect support to her grown offspring, the fact remains: individuals fare better when their mothers are around.

Observations of death (of either infants or mothers) allow us only brief glimpses of the emotional and psychological impacts that it has on the survivor. We often cannot tell how members of our own species are faring after they have experienced the death of a loved one with any great certainty, so how are we supposed to categorize and quantify the actions of our closest animal cousins? This is the inevitable challenge when scientific rigor meets emotional complexity. Both of these are at the heart of understanding the cognitive nuances of the mourning process; however, they are areas of research that don't often overlap. As our data sets grow to encompass more and more instances of mourning in the animal kingdom, we may find that certain generalizations are a good fit. Until then, we must recognize the universal truth that the mourning process is an individual phenomenon. The intersection of sadness, loss, and emotional irrationality against the backdrop of continued survival of the fittest is something that we may never completely understand.

AFTERWORD

My perspective on motherhood has changed dramatically during the course of researching and writing this book, which is somewhat fascinating since I have four children of my own and I've been therefore "walking the walk" for the last thirteen years. On some level, it seems that my biological and emotional selves have been subjected to maternal processes, whereas my philosophical self has lagged behind. In living the hectic day-to-day of motherhood, I didn't always have a moment to take the long view of it in our culture. It was a perspective that I gained in writing this book. For that, I will always be grateful.

Given this time to reflect from my twenty-thousand-foot view of motherhood, I thought about a lot of things. Time and time again, I was awe-inspired by the efficiency of our biological machinery and how much work females' bodies do in the construction of healthy infants without any kind of cognizance on the part of the mother experiencing it. When a mammalian female (human or otherwise) begins to gestate an embryo, a combination of irreversible physiological and psychological changes is set in motion. It's this irreversibility which I find most compelling, because a female is no longer the same animal, nor will she ever be again.

Human females are unique in the animal kingdom in the combination of our massive brains, low fecundity, and long lifespans, including several decades of being postreproductive. Our species is the only one in which mothers have agendas that don't always involve biology. This is largely because we have developed straightforward methods of securing our biological needs like food and shelter—no need for time-consuming hunting when we can just drive to the supermarket—and we have a very low risk of being attacked by predators. Because these needs are so easily met, we naturally have much more time left over to focus on other things—such as professional careers.

The irreversibility of the maternal transformation and the relative ease of our daily lives make me question the way in which human societies compartmentalize the process of motherhood. On a daily basis, women are expected to make difficult decisions between being powerful in their professional lives or being successful at motherhood. I think that this career conundrum places many human mothers in a situation that is unequalled in the animal kingdom—their brains prior to child rearing are used to being engaged with professional work that doesn't directly relate to that for which their "new mom" brains are biologically built. The discord between professional brain and mommy brain can make the transition into motherhood a slow and somewhat tumultuous one, as each mother aims for that ever-elusive balance between career and family. I would argue that several aspects of postpartum depression in career women are related to this abrupt neurophysiological shift in brain power from purely professional to purely biological. This is a dilemma faced by many women in Western societies, and although it's an individual-level conundrum in which new moms will strive to figure out an approximate balance that works for themselves and their families, there are inherent biases at all scales of the human social and professional construct.

This is where society in general gets it wrong. By devaluing the immense task of motherhood in the professional world, we place professional mothers on a lower rung of whatever career-ladder they may be climbing—which then leads to male-biased ratios of individuals in high positions of most professions. What much of the world has not yet grasped about the reconciliation of motherhood and professionalism is that mothers bring an entirely new suite of ideas, perspectives, and approaches that are naturally part of what their own bodies and minds have become. The biological changes that make us mothers include several characteristics that render us uniquely qualified to excel at our former positions in completely new ways.

This leads me to a critically important point about the function of motherhood in human societies: there is an immense lack of understanding the all-encompassing biological phenomenon that motherhood is. Career women who are also mothers should not be characterized as weak, distracted, or otherwise dismissible. I wonder how modern societies would differ if we were to value the biological contributions of women rather than considering them an interference to the so-called "important" work to be done in the world. What if we awarded career moms with the kind of support systems that they need in order to be successful at both? What if motherhood was actually celebrated in the career world instead of shunned? The whole world could change as a result.

If you've taken just one thing from reading this book, I hope it is that the enterprise of motherhood is far more than a simple output of offspring. Motherhood is an individual, enigmatic, and awe-inspiring journey, and there is simply no one right way to be a mom. It is perhaps the most important and meaningful process that evolution ever made. Have you called your mom lately to express your gratitude for her selfless feats? Maybe you should. She deserves it.

REFERENCES

1: DEVELOPMENT IS DIVERSE

Aron, S., Mardulyn, P., Leniaud, L. 2016. "Evolution of reproductive traits in Cataglyphis desert ants: mating frequency, queen number, and thelytoky." *Behavioral Ecology and Sociobiology* 70: 1367–1379.

Gibson, G.D., Gibson, A.J.F. 2004. "Heterochrony and the Evolution of Poecilogony: Generating Larval Diversity." *Evolution* 58(12): 2704–2717.

Ma, W.J., Schwander, T. 2017. "Patterns and mechanisms in instances of endosymbiont-induced parthenogenesis." *Journal of Evolutionary Biology* 30: 868–888.

Olito, C., Marshall, D.J., Conallon, T. 2016. "The Evolution of Reproductive Phenology in Broadcast Spawners and the Maintenance of Sexually Antagonistic Polymorphism." *The American Naturalist* 189(2): 153–169.

Sandberger-Loua, L., Müller, H., Rödel, M. 2017. "A review of the reproductive biology of the only known matrotrophic viviparous anuran, the West African Nimba toad, Nimbaphrynoides occidentalis." *Zoosystematics and Evolution* 93(1): 105–133.

Swift, D.G., et al. 2016. "Evidence of positive selection associated with placental loss in tiger sharks." *BMC Evolutionary Biology* 16: 126.

Velo-Antón, G., et al. 2014. "Intraspecific variation in clutch size and maternal investment in pueriparous and larviparous Salamandra salamandra females." *Evolutionary Ecology* 29: 185–204.

Walker, L.A., Holwell, G.I. 2015. "Sexual cannibalism in a facultative parthenogen: the springbok mantis (Miomantis caffra)." *Behavioral Ecology* 27(3): 851–856.

2: THE EGG-CONTENT CONUNDRUM

Beamonte-Barrientos, R., et al. 2010. "Senescence of Maternal Effects: Aging Influences Egg Quality and Rearing Capacities of a Long-Lived Bird." *The American Naturalist* 175(4): 469–480.

Benowitz-Fredericks, Z.M., et al. 2013. "Effects of Food Availability on Yolk Androgen Deposition in the Black-Legged Kittiwake (*Rissa tridactyla*), a Seabird with Facultative Brood Reduction." *PLoS ONE* 8(5): e62949.

Cunningham, E.J.A., Lewis, S. 2005. "Parasitism of maternal investment selects for increased clutch size and brood reduction in a host." *Behavioral Ecology* 17: 126–131.

Dentressangle, F., et al. 2008. "Maternal investment in eggs is affected by male feet colour and breeding conditions in the blue-footed booby, *Sula nebouxii*." *Behavioral Ecology and Sociobiology* 62: 1899–1908.

Drummond, H., et al. 2008. "Do mothers regulate facultative and obligate siblicide by differentially provisioning eggs with hormones?" *Journal of Avian Biology* 39: 139–143.

Giraudeau, M., Ducatez, S. 2016. "Co-adjustment of yolk antioxidants and androgens in birds." *Biology Letters* 2: 20160676.

Groothuis, T.G.G., et al. 2005. "Maternal hormones as a tool to adjust offspring phenotype in avian species." *Neuroscience and Biobehavioral Reviews* 29: 329–352.

Hsu, B., et al. 2015. "Maternal adjustment or constraint: differential effects of food availability on maternal deposition of macro-nutrients, steroids and thyroid hormones in rock pigeon eggs." *Ecology and Evolution* 6(2): 397–411.

Krištofík, J., et al. 2014. "Do females invest more into eggs when males sing more attractively? Postmating sexual selection strategies in a monogamous reed passerine." *Ecology and Evolution* 4(8): 1328–1339.

Lahaye, S.E.P., et al. 2015. "Influence of mate preference and laying order on maternal allocation in a monogamous parrot species with extreme hatching asynchrony." *Hormones and Behavior* 71: 49–59.

Langmore, N.E., et al. 2016. "Egg size investment in superb fairy-wrens: helper effects are modulated by climate." *Proceedings of the Royal Society* B. 283: 1–8.

Luque, S.P., et al. 2007. "Foraging behaviour of sympatric Antarctic and subantarctic fur seals: does their contrasting duration of lactation make a difference?" *Marine Biology* 152: 213–224.

Markham, A.C., et al. 2013. "Rank effects on social stress in lactating chimpanzees." *Animal Behavior* 87: 195–202.

McDonald, B.I., et al. 2012. "Dynamic Influence of Maternal and Pup Traits on Maternal Care during Lactation in an Income Breeder, the Antarctic Fur Seal." *Physiological and Biochemical Zoology* 85(3): 243–254.

Muller, M., Groothuis, T.G.G. 2013. "Within-Clutch Variation in Yolk Testosterone as an Adaptive Maternal Effect to Modulate Avian Sibling Competition: Evidence from a Comparative Study." *The American Naturalist* 181(1): 125–136.

Muller, M.S., et al. 2012. "Maternal Androgens Increase Sibling Aggression, Dominance, and Competitive Ability in the Siblicidal Black-Legged Kittiwake (*Rissa tridactyla*)." *PLoS ONE* 7(10): e47763.

Osthoff, G., et al. 2017. "Milk composition of free-ranging red hartebeest, giraffe, Southern reedbuck and warthog and a phylogenetic comparison of the milk of African Artiodactyla." *Comparative Biochemistry and Physiology*, Part A 204: 93–103.

Paquet, M., Smiseth, P.T. 2016. "Maternal effects as a mechanism for manipulating male care and resolving sexual conflict over care." *Behavioral Ecology* 27(3): 685–694.

Paquet, M. 2013. "Maternal Effects in Relation to Helper Presence in the Cooperatively Breeding Sociable Weaver." *PLoS ONE* 8(3): e59336.

Parolini, M., et al. 2015. "Vitamin E deficiency in last-laid eggs limits growth of yellow-legged gull chicks." *Functional Ecology* 29: 1070–1077.

Poisbleau, M., et al. 2012. "Penguin Chicks Benefit from Elevated Yolk Androgen Levels under Sibling Competition." *PLoS ONE* 7(7): e42174.

Ramírez, F., et al. 2015. "Older female little penguins *Eudyptula minor* adjust nutrient allocations to both eggs." *Journal of Experimental Marine Biology and Ecology* 468: 91–96.

Reed, W.L., et al. 2009. "Maternal Effects Increase Within-Family Variation in Offspring Survival." *The American Naturalist* 174(5): 685–695.

Reich, C. M., Arnould, J.P.Y. 2007. "Evolution of Pinnipedia lactation strategies: a potential role for α-lactalbumin?" *Biology Letters* 3: 546–549.

Remeš, V. 2011. "Yolk androgens in great tit eggs are related to male attractiveness, breeding density and territory quality." *Behavioral Ecology and Sociobiology* 65(6): 1257–1266.

Saino, N., et al. 2010. "Food supplementation affects egg albumen content and body size asymmetry among yellow-legged gull siblings." *Behavioral Ecology and Sociobiology* 64(11): 1813–1821.

Saino, N., et al. 2011. "Hatching asynchrony and offspring sex: an experiment on maternal effects in the yellow-legged gull." *Ethology Ecology & Evolution* 23: 4, 300–317.

Thompson, M.E., et al. 2012. "The energetics of lactation and the return to fecundity in wild chimpanzees." *Behavioral Ecology* 23: 1234–1241.

3: BROOD PARASITISM: AN INCONVENIENT TRUTH

Barton, R.A., et al. 2011. "Maternal investment life histories, and the costs of brain growth in mammals." *Proceedings of the National Academy of Sciences of the United States of America* 108(15): 6169–6174.

Brooker, M., Brooker, L. 1995. "Acceptance by the splendid fairy-wren of parasitism by Horsfield's bronze-cuckoo: further evidence for evolutionary equilibrium in brood parasitism." *Behavioral Ecology* 7(4): 395–407.

Elliot, M.G., Crespi, B.J. 2008. "Placental invasiveness and brain–body allometry in eutherian Mammals." *Journal of Evolutionary Biology* 21: 1763–1778.

Feeney, W.E., et al. 2012. "The frontline of avian brood parasite-host coevolution." *Animal Behaviour* 84: 3–12.

Fiorini, V.D., et al. 2014. "Strategic egg destruction by brood-parasitic cowbirds?" *Animal Behaviour* 93: 229–235.

Gloag, R., et al. 2013. "The wages of violence: mobbing by mockingbirds as a frontline defence against brood-parasitic cowbirds." *Animal Behaviour* 86: 1023–1029.

Grim, T., et al. 2011. "Constraints on host choice: why do parasitic birds rarely exploit some common potential hosts?" *Journal of Animal Ecology* 80: 508–511.

Jelínek, V., et al. 2016. "Do common cuckoo chicks suffer nest predation more than host nestlings?" *Behavioral Ecology and Sociobiology* 70: 1975–1987.

Langmore, N.E., et al. 2003. "Escalation of a coevolutionary arms race through host rejection of brood parasitic young." *Nature* 422: 157–160.

Lewitus, E., Kalinka, A.T. 2013. "Neocortical development as an evolutionary platform for intragenomic conflict." *Frontiers in Neuroanatomy* 7: 1–13.

Lichtenstein, G. 2001. "Low success of shiny cowbird chicks parasitizing rufous-bellied thrushes: chick–chick competition or parental discrimination?" *Animal Behaviour* 61: 401–413.

Martin. R.D. 2007. "The Evolution of Human Reproduction: A Primatological Perspective." *Yearbook of Physical Anthropology* 50: 59–84.

Massoni, V., Reboreda, J.C. 1998. "Costs of brood parasitism and the lack of defenses on the yellow-winged blackbird-shiny cowbird system." *Behavioral Ecology and Sociobiology* 42: 273–280.

Morton, E.S., Farabaugh, S.M. 1979. "Infanticide and Other Adaptions of the Nestling Striped Cuckoo *Tapera Naevia*." *Ibis* 121: 212–213.

Remeš, V. 2006. "Growth strategies of passerine birds are related to brood parasitism by the brown-headed cowbird (*Molothrus ater*)." *Evolution* 60(8): 1692–1700.

Rivers, J.W., et al. 2010. "Brown-headed cowbird nestlings influence nestmate begging, but not parental feeding, in hosts of three distinct sizes." *Animal Behaviour* 79: 107–116.

Sato, T. 1986. "A brood parasitic catfish of mouthbrooding cichlid fishes in Lake Tanganyika." *Nature* 323(4): 58–59.

Scardamaglia, R.C., et al. 2017. "Planning host exploitation through prospecting visits by parasitic cowbirds." *Behavioral Ecology and Sociobiology* 71: 23.

Sealy, S.G. 1995. "Burial of cowbird eggs by parasitized yellow warblers: an empirical and experimental study." *Animal Behaviour* 49: 877–889.

Soler, J.J., et al. 1995. "Does the great spotted cuckoo choose magpie hosts according to their parenting ability?" *Behavioral Ecology and Sociobiology* 36: 201–206.

Soler, M. 2014. "Long-term coevolution between avian brood parasites and their hosts." *Biology Reviews* 89: 688–704.

Soler, M., et al. 2014. "Great Spotted Cuckoos Frequently Lay Their Eggs While Their Magpie Host Is Incubating." *Ethology* 120: 965–972.

Soler, M., Pérez-Contreras, T. 2012. "Location of suitable nests by great spotted cuckoos: an empirical and experimental study." *Behavioral Ecology and Sociobiology* 66(9): 1305–1310.

Soler, M., et al. 2015. "Synchronization of laying by great spotted cuckoos and recognition ability of magpies." *Journal of Avian Biology* 46: 608–615.

Spottiswoode, C.N., Stevens, M. 2011. "How to evade a coevolving brood parasite: egg discrimination versus egg variability as host defences." *Proceedings of the Royal Society B* 278: 3566–3573.

Stoddard, M.C., Stevens, M. 2011. "Avian Vision and the Evolution of Egg Color Mimicry In the Common Cuckoo." *Evolution* 65(7): 2004–2013.

4: COOPERATIVE CREEDS AND COMMUNAL CLIQUES

Baden, A.L., et al. 2013. "Communal nesting, kinship, and maternal success in a social primate." *Behavioral Ecology and Sociobiology* 67(12): 1939–1950.

Bădescu, I., et al. 2016. "Infanticide pressure accelerates infant development in a wild primate." *Animal Behaviour* 114: 231–239.

Bauer, C.M., et al. 2015. "Maternal stress and plural breeding with communal care affect development of the endocrine stress response in a wild rodent." *Hormones and Behavior* 75: 18–24.

Bijleveld, A.I., et al. 2009. "Beyond the information centre hypothesis: communal roosting for information on food, predators, travel companions and mates?" *Oikos* 119: 277285.

Brunelli, S.A. 2005. "Development and Evolution of Hidden Regulators: Selective Breeding for an Infantile Phenotype." *Developmental Psychobiology* 47: 243–252.

Burkart, J.M., Van Schaik, C.P. 2015. "Revisiting the consequences of cooperative breeding." *Journal of Zoology* 299: 77–83.

Ciszek, D. 2000. "New colony formation in the 'highly inbred' eusocial naked mole-rat: outbreeding is preferred." *Behavioral Ecology* 11(1): 1–6.

Clarke, F.M., Faulkest, C.G. 2000. "Intracolony aggression in the eusocial naked mole-rat, *Heterocephalus glaber.*" *Animal Behaviour* 61: 311–324.

Clutton-Brock, T.H., et al. 2006. "Intrasexual competition and sexual selection in cooperative mammals." *Nature* 444: 1065–1068.

Clutton-Brock, T.H., Huchard, E. 2013. "Social competition and its consequences in female mammals." *Journal of Zoology* 289: 151–171.

Culot, L., et al. 2011. "Reproductive failure, possible maternal infanticide, and cannibalism in wild moustached tamarins, *Saguinus mystax.*" *Primates* 52: 179–186.

DeCatanzaro, D. 2015. "Sex steroids as pheromones in mammals: The exceptional role of estradiol." *Hormones and Behavior* 68: 103–116.

Díaz-Muñoz, S.L. 2016. "Complex Cooperative Breeders: Using Infant Care Costs to Explain Variability in Callitrichine Social and Reproductive Behavior." *American Journal of Primatology* 78: 372–387.

Digby, L. 1995. "Infant care, infanticide, and female reproductive strategies in polygynous groups of common marmosets (*Callithrix jacchus*)." *Behavioral Ecology and Sociobiology* 37: 51–61.

Digby, L. 1999. "Sexual Behavior and Extragroup Copulations in a Wild Population of Common Marmosets (*Callithrix jacchus*)." *Folia Primatol* 70: 136–145.

Eberle, M., Kappeler, P.M. 2006. "Family Insurance: Kin Selection and Cooperative Breeding in a Solitary Primate (Microcebus murinus)." *Behavioral Ecology and Sociobiology* 60(4): 582–588.

Ebneter, C., et al. 2016. "A trade-off between reproductive investment and maternal cerebellum size in a precocial bird." *Biology Letters* 12: 20160659.

Erb, W.M., Porter, L.M. 2017. "Mother's little helpers: What we know (and don't know) about cooperative infant care in callitrichines." *Evolutionary Anthropology* 26: 25–37.

Ferrari, M., et al. 2015. "The risk of exploitation during communal nursing in house mice, *Mus musculus domesticus.*" *Animal Behaviour* 110: 133–143.

Guzzo, A.C., et al. 2012. "Oestradiol transmission from males to females in the context of the Bruce and Vandenbergh effects in mice (*Mus musculus*)." *Reproduction* 143: 539–548.

Hathaway, G.A., et al. 2016. "Subcaste differences in neural activation suggest a prosocial role for oxytocin in eusocial naked mole-rats." *Hormones and Behavior* 79: 1–7.

Hayes, L.D., Solomon, N.G. 2004. "Costs and Benefits of Communal Rearing to Female Prairie Voles (Microtus ochrogaster)." *Behavioral Ecology and Sociobiology* 56(6): 585–593.

REFERENCES

Henry, M.D., et al. 2013. "High rates of pregnancy loss by subordinates leads to high reproductive skew in wild golden lion tamarins (*Leontopithecus rosalia*)." *Hormones and Behavior* 63: 675–683.

Hill, D.L., et al. 2015. "Alternative reproductive tactics in female striped mice: heavier females are more likely to breed solitarily than communally." *Journal of Animal Ecology* 84: 1497–1508.

Hodge, S.J., et al. 2008. "Determinants of reproductive success in dominant female meerkats." *Journal of Animal Ecology* 77: 92–102.

Holmes, S.M., et al. 2016. "Fission-fusion dynamics in black-and-white ruffed lemurs may facilitate both feeding strategies and communal care of infants in a spatially and temporally variable environment." *Behavioral Ecology and Sociobiology* 70: 1949–1960.

Hood, W.R., et al. 2014. "Milk composition and lactation strategy of a eusocial mammal, the naked mole-rat." *Journal of Zoology* 293: 108–118.

Hoogland, J.L. 2009. "Nursing of Own and Foster Offspring by Utah Prairie Dogs (Cynomys parvidens)." *Behavioral Ecology and Sociobiology* 63(11): 1621–1634.

Hrdy, S.B. 2016. "Variable postpartum responsiveness among humans and other primates with 'cooperative breeding': A comparative and evolutionary perspective." *Hormones and Behavior* 77: 272–283.

Jaggard, A.K., et al. 2015. "Rules of the roost: characteristics of nocturnal communal roosts of rainbow lorikeets (*Trichoglossus haematodus, Psittacidae*) in an urban environment." *Urban Ecosystems* 18: 489–502.

Kerth, G., et al. 2001. "Day roost selection in female Bechstein's bats (Myotis bechsteinii): a field experiment to determine the influence of roost temperature." *Oecologia* 126:1–9.

Kleindorfer, S., Wasser, S.K. 2004. "Infant Handling and Mortality in Yellow Baboons (*Papio cynocephalus*): Evidence for Female Reproductive Competition?" *Behavioral Ecology and Sociobiology* 56(4): 328–337.

Kramer, K.L., Otárola-Castillo, E. 2015. "When mothers need others: The impact of hominin life history evolution on cooperative breeding." *Journal of Human Evolution* 84: 16–24.

Kutsukake, N., Clutton-Brock, T.H. 2006. "Aggression and submission reflect reproductive conflict between females in cooperatively breeding meerkats *Suricata suricatta*." *Behavioral Ecology and Sociobiology* 59: 541–548.

Lambertucci, S.A., Ruggiero, A. 2016. "Cliff outcrops used as condor communal roosts are local hotspots of occupancy and intense use by other bird species." *Biological Conservation* 200: 8–16.

Lausen, C.L., Barclay, R.M.R. "Thermoregulation and roost selection by reproductive female big brown bats (*Eptesicus fuscus*) roosting in rock crevices." *Journal of Zoology* 260: 235–244.

MacLeod, K.J., Clutton-Brock, T.H. 2015. "Low costs of allonursing in meerkats: mitigation by behavioral change?" *Behavioral Ecology* 26(3): 697–705.

MacLeod, K.J., et al. 2015. "No apparent benefits of allonursing for recipient offspring and mothers in the cooperatively breeding meerkat." *Journal of Animal Ecology* 84: 1050–1058.

Maniscalco, J.M., et al. 2007. "Alloparenting in Steller sea lions (*Eumetopias jubatus*): correlations with misdirected care and other observations." *Journal of Ethology* 25: 125–131.

McConville, A., et al. 2013. "Mangroves as maternity roosts for a colony of the rare east-coast free-tailed bat (*Mormopterus norfolkensis*) in south-eastern Australia." *Wildlife Research* 40: 318–327.

Mucignat-Caretta, C., Caretta, A. 2014. "Message in a bottle: major urinary proteins and their multiple roles in mouse intraspecific chemical communication." *Animal Behaviour* 97: 255–263.

Nelson-Flower, M.J., et al. 2013. "Costly reproductive competition between females in a monogamous cooperatively breeding bird." *Proceedings of the Royal Society B* 280: 20130728.

Pusey, A.E., Packer, C. 1993. "Non-offspring nursing in social carnivores: minimizing the costs." *Behavioral Ecology* 5(4): 362–374.

Roberts, E.K., et al. 2012. "A Bruce Effect in Wild Geladas." *Science* 335(6073): 1222–1225.

Rohwer, S., et al. 1999. "Stepparental Behavior as Mating Effort in Birds and Other Animals." *Evolution and Human Behavior* 20: 367–390.

Roulin, A., Hager, R. 2003. "Indiscriminate nursing in communal breeders: a role for genomic imprinting." *Ecology Letters* 6: 165–166.

Saltzman, W., et al. 2004. "Onset of plural cooperative breeding in common marmoset families following replacement of the breeding male." *Animal Behaviour* 68: 59–73.

Schmidt, J., et al. 2015. "Reproductive asynchrony and infanticide in house mice breeding communally." *Animal Behaviour* 101: 201–211.

Sharpe, L.L., et al. 2016. "Robbing rivals: interference foraging competition reflects female reproductive competition in a cooperative mammal." *Animal Behaviour* 112: 229–236.

Stanton, M.A. 2015. "Maternal Behavior and Physiological Stress Levels in Wild Chimpanzees (*Pan troglodytes schweinfurthii*)." *International Journal of Primatology* 6: 473–488.

Stanton, M.A., et al. 2014. "Maternal Behavior by Birth Order in Wild Chimpanzees (*Pan troglodytes*)." *Current Anthropology* 55: 483–489.

St. Germain, M.J., et al. 2017. "Who Knew? First *Myotis sodalis* (Indiana Bat) Maternity Colony in the Coastal Plain of Virginia." *Northeastern Naturalist*, 24(1):N5–N10.

Tecot, S.R., et al. 2012. "Infant parking and nesting, not allomaternal care, influence Malagasy primate life histories." *Behavioral Ecology and Sociobiology* 66(10): 1375–1386.

Thavarajah, N.K., et al. 2013. "The determinants of dominance relationships among subordinate females in the cooperatively breeding meerkat." *Behaviour* 151: 89–102.

Toor, I., et al. 2015. "Olfaction and social cognition in eusocial naked mole-rats, *Heterocephalus glaber*." *Animal Behaviour* 107: 175–181.

Toth, C.A., et al. 2015. "Females as mobile resources: communal roosts promote the adoption of lek breeding in a temperate bat." *Behavioral Ecology* 26(4): 1156–1163.

Townsend, S.W., et al. 2007. "Female-led infanticide in wild chimpanzees." *Current Biology* 17(10): R355–R356.

Wilkinson, G.S. 1992. "Communal nursing in the evening bat, *Nycticeius humeralis*." *Behavioral Ecology and Sociobiology* 31: 225–235.

Williams, S.A., Shattuck, M.R. 2014. "Ecology, longevity and naked mole-rats: confounding effects of sociality?" *Proceedings of the Royal Society B* 282: 20141664.

Yamagiwa, J., et al. 2009. "Infanticide and social flexibility in the genus Gorilla." *Primates* 50: 293–303.

Ziegler, T.E. 2013. "Social Effects via Olfactory Sensory Stimuli on Reproductive Function and Dysfunction in Cooperative Breeding Marmosets and Tamarins." *American Journal of Primatology* 75: 202–211.

5: BUNS IN MAMMALIAN OVENS

Angiolini, E., et al. 2006. "Regulation of Placental Efficiency for Nutrient Transport by Imprinted Genes." *Placenta* 27: Supplement A, Trophoblast Research, Vol. 20.

Bailey, A., et al. 2015. "Does pregnancy coloration reduce female conspecific aggression in the presence of maternal kin?" *Animal Behaviour* 108: 199–206.

Beehner, J.C., et al. 2006. "The ecology of conception and pregnancy failure in wild baboons." *Behavioral Ecology* 17: 741–750.

Beehner, J.C., Lu, A. 2013. "Reproductive Suppression in Female Primates: A Review." *Evolutionary Anthropology* 22: 226–238.

Bercovitch, F.B., et al. 2000. "Maternal investment in rhesus macaques (*Macaca mulatta*): reproductive costs and consequences of raising sons." *Behavioral Ecology and Sociobiology* 48: 1–11.

Berghänel, A., et al. 2016. "Prenatal stress effects in a wild, long-lived primate: predictive adaptive responses in an unpredictable environment." *Proceedings of the Royal Society B* 283: 20161304.

Bressan, F.F., et al. 2009. "Unearthing the Roles of Imprinted Genes in the Placenta." *Placenta* 30: 823–834.

Capellini, I. 2012. "The evolutionary significance of placental interdigitation in mammalian reproduction: Contributions from comparative studies." *Placenta* 33: 763–768.

Capellini, I., et al. 2015. "Microparasites and Placental Invasiveness in Eutherian Mammals." *PLoS ONE* 10(7): e0132563.

Capellini, I., et al. 2011. "Placentation and Maternal Investment in Mammals." *The American Naturalist* 177(1): 86–98.

Carter, A.M., Enders, A.C. 2016. "Placentation in mammals: Definitive placenta, yolk sac, and paraplacenta." *Theriogenology* 86: 278–287.

Chavan, A.R., et al. 2016. "What was the ancestral function of decidual stromal cells? A model for the evolution of eutherian pregnancy." *Placenta* 40: 40–51.

Constância, M., et al. 2004. "Resourceful imprinting." *Nature* 432: 53–57.

Crawford, J.C., Drea, C.M. 2015. "Baby on board: olfactory cues indicate pregnancy and fetal sex in a non-human primate." *Biology Letters* 11: 20140831.

Domínguez, M., et al. 2015. "Host switching in cowbird brood parasites: how often does it occur?" *Journal of Evolutionary Biology* 28: 1290–1297.

Douglas, P.H., et al. 2016. "Mixed messages: wild female bonobos show high variability in the timing of ovulation in relation to sexual swelling patterns." *BMC Evolutionary Biology* 16: 140.

Enjapoori, A.K., et al. 2014. "Monotreme Lactation Protein Is Highly Expressed in Monotreme Milk and Provides Antimicrobial Protection." *Genome Biology and Evolution* 6(10): 2754–2773.

Fitzpatrick, C.L., et al. 2015. "Exaggerated sexual swellings and male mate choice in primates: testing the reliable indicator hypothesis in the Amboseli baboons." *Animal Behaviour* 104: 175–185.

Fowden, A.L., et al. 2011. "Imprinted genes and the epigenetic regulation of placental phenotype." *Progress in Biophysics and Molecular Biology* 106: 281–288.

Fowden, A.L., Moore, T. 2012. "Maternal-fetal resource allocation: Co-operation and conflict." *Placenta* 33: e11–e15.

Garratt, M., et al. "Diversification of the eutherian placenta is associated with changes in the pace of life." *Proceedings of the National Academy of Sciences of the United States of America* 110(19): 7760–7765.

Georges, J., Guinet, C. 2000. "Maternal Care in the Subantarctic Fur Seals on Amsterdam Island." *Ecology* 81(2): 295–308.

Grützner, F., et al. 2008. "Reproductive Biology in Egg-Laying Mammals." *Sexual Development* 2: 115–127.

Gundling, W.E., Wildman, D.E. 2015. "A review of inter- and intraspecific variation in the eutherian placenta." *Philosophical Transactions of the Royal Society B* 370: 20140072.

Hackländer, K., Arnold, W. 1999. "Male-caused failure of female reproduction and its adaptive value in alpine marmots (*Marmota marmota*)." *Behavioral Ecology* 10: 592–597.

Holand, Ø., et al. 2006. "Adaptive Adjustment of Offspring Sex Ratio and Maternal Reproductive Effort in an Iteroparous Mammal." *Proceedings: Biological Sciences* 273(1584): 293–299.

Lahdenperä, M., et al. 2016. "Short-term and delayed effects of mother death on calf mortality in Asian elephants." *Behavioral Ecology* 27(1): 166–174.

Lee, P.C., et al. 2016. "The reproductive advantages of a long life: longevity and senescence in wild female African elephants." *Behavioral Ecology and Sociobiology* 70: 337–345.

Lefevre, C.M., et al. 2010. "Evolution of Lactation: Ancient Origin and Extreme Adaptations of the Lactation System." *Annual Review of Genomics and Human Genetics* 11: 219–38.

Lueders, I., et al. 2012. "Gestating for 22 months: luteal development and pregnancy maintenance in elephants." *Proceedings of the Royal Society B* 279(1743): 3687–3696.

Maestripieri, D. 1999. "Changes in Social Behavior and Their Hormonal Correlates during Pregnancy in Pig-tailed Macaques." *International Journal of Primatology* 20(5): 707–718.

Martin, J.G.A., Festa-Bianchet, M. 2011. "Determinants and consequences of age of primiparity in bighorn ewes." *Oikos* 121: 752–760.

Moore, T. 2012. "Review: Parent-offspring conflict and the control of placental function." *Placenta* 33, Supplement A, Trophoblast Research, Vol. 26: 533–536.

Morrow, G.E., Nicol, S.C. 2012. "Maternal care in the Tasmanian echidna (*Tachyglossus aculeatus setosus*)." *Australian Journal of Zoology* 60(5): 289–298.

Ng, H.K., et al. 2010. "Distinct Patterns of Gene-Specific Methylation in Mammalian Placentas: Implications for Placental Evolution and Function." *Placenta* 31: 259–268.

Ryan, S.J., et al. 2007. "Ecological cues, gestation length, and birth timing in African buffalo (*Syncerus caffer*)." *Behavioral Ecology* 18: 635–644.

Temple-Smith, P., Grant, T. 2001. "Uncertain breeding: a short history of reproduction in monotremes." *Reproducation, Fertility and Development* 13: 487–497.

Thompson, M.E. 2013. "Comparative Reproductive Energetics of Human and Non-human Primates." *Annual Review of Anthropology* 42: 287–304.

Thompson, M.E. 2013. "Reproductive Ecology of Female Chimpanzees." *American Journal of Primatology* 75: 222–237.

Van Dierendonck, M.C., et al. 2004. "Differences in social behaviour between late pregnant, post-partum and barren mares in a herd of Icelandic horses." *Applied Animal Behaviour Science* 89: 283–297.

Wang, X., et al. 2013. "Paternally expressed genes predominate in the placenta." *Proceedings of the National Academy of Sciences of the United States of America* 110(26): 10705–10710.

6: A PREGNANT PAUSE

Fenelon, J.C., et al. 2014. "Embryonic diapause: development on hold." *International Journal of Developomental Biology* 58: 163–174.

Fenelon, J.C., et al. 2017. "Regulation of diapause in carnivores." *Reproduction in Domestic Animals* 52: 12–17.

Orr, T.J., Zuk, M. 2014. "Reproductive delays in mammals: an unexplored avenue for post-copulatory sexual selection." *Biology Reviews* 89: 889–912.

Ptak, G.E., et al. 2013. "Embryonic diapause in humans: time to consider?" *Reproductive Biology and Endocrinology* 11: 92.

Ptak, G.E., et al. 2012. "Embryonic Diapause Is Conserved across Mammals." *PLoS ONE* 7(3): e33027.

Renfree, M.B. 2015. "Embryonic Diapause and Maternal Recognition of Pregnancy in Diapausing Mammals." *Advances in Anatomy Embryology and Cell Biology* 216: 239–252.

7: A CHILD IS BORN

Ancillotto, L., et al. 2015. "Sociality across species: spatial proximity of newborn bats promotes heterospecific social bonding." *Behavioral Ecology* 26(1): 293–299.

Bautista, A., et al. 2015. "Contribution of within-litter interactions to individual differences in early postnatal growth in the domestic rabbit." *Animal Behaviour* 108: 145–153.

Bautista, A., et al. 2015. "Intrauterine position as a predictor of postnatal growth and survival in the rabbit." *Physiology & Behaviour* 138: 101–106.

Culot, L., et al. 2011. "Reproductive failure, possible maternal infanticide, and cannibalism in wild moustached tamarins, *Saguinus mystax*." *Primates* 52: 179–186.

Douglas, P.H. 2014. "Female sociality during the daytime birth of a wild bonobo at Luikotale, Democratic Republic of the Congo." *Primates* 55: 533–542.

Dunsworth, H., Eccleston, L. 2015. "The Evolution of Difficult Childbirth and Helpless Hominin Infants." *Annual Review of Anthropology* 44: 55–69.

Ernest, S.K.M. 2003. "Life history characteristics of placental nonvolant mammals." *Ecology* 84(12): 3402.

Fujisawa, M., et al. 2016. "Placentophagy in wild chimpanzees (*Pan troglodytes verus*) at Bossou, Guinea." *Primates* 57: 175–180.

González-Mariscal, G., et al. 2016. "Mothers and offspring: The rabbit as a model system in the study of mammalian maternal behavior and sibling interactions." *Hormones and Behavior* 77: 30–41.

Haim, A., et al. 1992. "Burrowing and Huddling in Newborn Porcupine: The Effect on Thermoregulation." *Physiology & Behavior* 52: 247–250.

Hirata, S., et al. 2011. "Mechanism of birth in chimpanzees: humans are not unique among primates." *Biology Letters* 7: 686–688.

Lévy, F., et al. 2004. "Olfactory regulation of maternal behavior in mammals." *Hormones and Behavior* 46: 284–302.

Lingle, S., Riede, T. 2014. "Deer Mothers Are Sensitive to Infant Distress Vocalizations of Diverse Mammalian Species." *The American Naturalist* 184(4): 510–522.

Mitteroecker, P., et al. 2016. "Cliff-edge model of obstetric selection in humans." *PNAS* 113: 14680–14685.

Nadler, R.D. 1989. "Sexual Initiation in Wild Mountain Gorillas." *International Journal of Primatology* 10(2): 81–92.

Nguyen, N., et al. 2012. "Sex differences in the mother-neonate relationship in wild baboons: social, experiential and hormonal correlates." *Animal Behaviour* 83: 891–903.

Pan, W., et al. 2014. "Birth intervention and non-maternal infant-handling during parturition in a nonhuman primate." *Primates* 55: 483–488.

Paul, A., et al. 1993. "Reproductive Senescence and Terminal Investment in Female Barbary Macaques (*Macaca sylvanus*) at Salem." *International Journal of Primatology* 14(1): 105–124.

Ross, A.C., et al., 2010. "Maternal care and infant development in *Callimico goeldii* and *Callithrix jacchus*." *Primates* 51: 315–325.

Schaal, B. 2014. "Chemical signals 'selected for' newborns in mammals." *Animal Behaviour* 97: 289–299.

Scheiber, I.B.R., et al. 2017. "The importance of the altricial–precocial spectrum for social complexity in mammals and birds—a review." *Frontiers in Zoology* 14: 3.

Sibiryakova, O.V., et al. 2017. "Remarkable vocal identity in wild-living mother and neonate saiga antelopes: a specialization for breeding in huge aggregations?" *Science and Nature* 104: 11.

Sibly, R.M., Brown, J.H. 2009. "Mammal Reproductive Strategies Driven by Offspring Mortality-Size Relationships." *The American Naturalist* 173(6): E185–E199.

Silk, J.B. 1986. "Eating for Two: Behavioral and Environmental Correlates of Gestation Length Among Free-Ranging Baboons (*Papio cynocephalus*)." *International Journal of Primatology* 7(6): 583–602.

Thomsen, R., Soltis, J. 2000. "Socioecological Context of Parturition in Wild Japanese Macaques (*Macaca fuscata*) on Yakushima Island." *International Journal of Primatology* 21(4): 685–696.

Timmermans, P.J.A., Vossen, J.M.H. 1996. "The Influence of Repeated Motherhood on Periparturitional Behavior in Cynomolgus Macaques (*Macaca fascicularis*)." *International Journal of Psychology* 17(2): 277–296.

Torriani, M.V.G., et al. 2006. "Mother-Young Recognition in an Ungulate Hider Species: A Unidirectional Process." *The American Naturalist* 168(3): 412–419.

Trevathan, W. 2015. "Primate pelvic anatomy and implications for birth." *Philosophical Transactions of the Royal Society B* 370: 20140065.

Turner, S.E., et al. 2010. "Birth in Free-ranging *Macaca fuscata*." *International Journal of Primatology* 31: 15–17.

Yang, B., et al. 2016. "Daytime birth and postbirth behavior of wild *Rhinopithecus roxellana* in the Qinling Mountains of China." *Primates* 57: 155–160.

Yao, M., et al. 2012. "Parturitions in Wild White-Headed Langurs (*Trachypithecus leucocephalus*) in the Nongguan Hills, China." *International Journal of Primatology* 33: 888–904.

8: NEWBORNS AND NIGHTMARES

Ernest, S.K.M. 2003. "Life history characteristics of placental nonvolant mammal." *Ecology* 84(12): 3402.

González-Mariscal, G., et al. 2016. "Mothers and offspring: The rabbit as a model system in the study of mammalian maternal behavior and sibling interactions." *Hormones and Behavior* 77: 30-41.

Haim, A., et al. 1992. "Burrowing and Huddling in Newborn Porcupine: The Effect on Thermoregulation." *Physiology & Behavior* 52: 47-250.

Hirata, S., et al. 2011. "Mechanism of birth in chimpanzees: humans are not unique among primates." *Biology Letters* 7: 686–688.

Lévy, F., et al. 2004. "Olfactory regulation of maternal behavior in mammals." *Hormones and Behavior* 46: 284—302.

Lingle, S., Riede, T. 2014. "Deer Mothers Are Sensitive to Infant Distress Vocalizations of Diverse Mammalian Species." *The American Naturalist* 184(4): 510–522.

Mann, J., Smuts, B. 1999. "Behavioral Development in Wild Bottlenose Dolphin Newborns (*Tursiops sp.*)." *Behaviour* 136: 529-566.

Nguyen, N., et al. 2012. "Sex differences in the mother-neonate relationship in wild baboons: social, experiential and hormonal correlates." *Animal Behaviour* 83: 891–903.

Schaal, B. 2014. "Chemical signals 'selected for' newborns in mammals." *Animal Behaviour* 97: 289-299.

Scheiber, I.B.R., et al. 2017. "The importance of the altricial-precocial spectrum for social complexity in mammals and birds—a review." *Frontiers in Zoology* 14: 3.

Sibiryakova, O.V., et al. 2017. "Remarkable vocal identity in wild-living mother and neonate saiga antelopes: a specialization for breeding in huge aggregations?" *Science and Nature* 104: 11.

Sibly, R.M., Brown, J.H. 2009. "Mammal Reproductive Strategies Driven by Offspring Mortality-Size Relationships." *The American Naturalist* 173(6): E185-E199.

Torriani, M.V.G., et al. 2006. "Mother-Young Recognition in an Ungulate Hider Species: A Unidirectional Process." *The American Naturalist* 168(3): 412-419.

Weir, J.S., et al. 2008. "Dusky dolphin (*Lagenorhynchus obscurus*) subgroup distribution: are shallow waters a refuge for nursery groups?" *Canadian Journal of Zoology* 86: 1225-1234.

9: DEPRESSION AND ABUSE: NOT JUST A HUMAN PREDICAMENT

Agrati, D., Lonstein, J.S. 2016. "Affective changes during the postpartum period: Influences of genetic and experiential factors." *Hormones and Behavior* 77: 141–152.

Bardi, M., et al. 2004. "The Role of the Endocrine System in Baboon Maternal Behavior." *Biological Psychiatry* 55: 724–732.

Brent, L., et al. 2002. "Abnormal, Abusive, and Stress-Related Behaviors in Baboon Mothers." *Biological Psychiatry* 52: 1047–1056.

Cox, J.L. 1994. "Postnatal Depression in Primate Mothers: A Human Problem?" 3rd Schultz-Biegert Symposium, Kartause, Ittingen, Switzerland.

Dettmer, A.M., et al. 2016. "Associations between early life experience, chronic HPA axis activity, and adult social rank in rhesus monkeys." *Social Neuroscience* 12(1): 92–101.

Dettmer, A.M., et al. 2016. "Neonatal face-to-face interactions promote later social behaviour in infant rhesus monkeys." *Nature Communications* 7: 11940.

Drury, S.S., et al. 2016. "When mothering goes awry: Challenges and opportunities for utilizing evidence across rodent, nonhuman primate and human studies to better define the biological consequences of negative early caregiving." *Hormones and Behavior* 77: 182–192.

Feldman, R. 2016. "The neurobiology of mammalian parenting and the biosocial context of human caregiving." *Hormones and Behavior* 77: 3–17.

French, J.A., Carp, S.B. 2016. "Early-life social adversity and developmental processes in nonhuman primates." *Current Opinion in Behavioural Sciences* 7: 40–46.

Graves, F.C., et al. 2002. "Opioids and Attachment in Rhesus Macaque (*Macaca mulatta*) Abusive Mothers." *Behavioural Neuroscience* 116(3): 489–493.

Hillerer, K.M., et al. 2012. "From Stress to Postpartum Mood and Anxiety Disorders: How Chronic Peripartum Stress Can Impair Maternal Adaptations." *Neuroendocrinology* 95: 22–38.

Howell, B.R., et al. 2014. "Early Adverse Experience Increases Emotional Reactivity in Juvenile Rhesus Macaques: Relation to Amygdala Volume." *Developmental Psychobiology* 56(8): 1735–1746.

Kim, S., et al. 2014. "Maternal oxytocin response predicts mother-to-infant gaze." *Brain Research* 1580: 133–142.

Machin, A.J., Dunbar, R.I.M. 2011. "The brain opioid theory of social attachment: a review of the evidence." *Behaviour* 148: 985–1025.

Maestripieri, D. 1999. "Fatal Attraction: Interest in Infants and Infant Abuse in Rhesus Macaques." *American Journal of Physical Anthropology* 110: 17–25.

Maestripieri, D. 2011. "Emotions, Stress, and Maternal Motivation in Primates." *American Journal of Primatology* 73: 516–529.

Maestripieri, D., et al. 2000. "Adoption and maltreatment of foster infants by rhesus macaque abusive mothers." *Developmental Science* 3: 3, 287–293.

Maestripieri, D., Carroll, K.A. 1998. "Behavioral and environmental correlates of infant abuse in group-living pigtail macaques." *Infant Behavior and Development* 21(4): 603–612.

Maestripieri, D., Carroll, K.A. 1998. "Risk Factors for Infant Abuse and Neglect in Group-Living Rhesus Monkeys." *Psychological Science* 9(2): 143–145.

Maestripieri, D., Carroll, K.A. 2000. "Causes and consequences of infant abuse and neglect in monkeys." *Aggression and Violent Behaviour* 5(3): 245–254.

Maestripieri, D., et al. 2000. "Crying and Infant Abuse in Rhesus Monkeys." *Child Development* 71(2): 301–309.

REFERENCES

Maestripieri, D., et al. 2009. "Mother–infant interactions in free-ranging rhesus macaques: Relationships between physiological and behavioral variables." *Physiology & Behavior* 96: 613–619.

Maestripieri, D., Maccoby, E.E. 2005. "Early Experience Affects the Intergenerational Transmission of Infant Abuse in Rhesus Monkeys." *Proceedings of the National Academy of Sciences of the United States of America* 102(27): 9726–9279.

Maestripieri, D., Megna, N.L. 2000. "Hormones and behavior in rhesus macaque abusive and nonabusive mothers 1. Social interactions during late pregnancy and early lactation." *Physiology & Behavior* 71: 35–42.

Maestripieri, D., Megna, N.L. 2000. "Hormones and behavior in rhesus macaque abusive and nonabusive mothers. 2. Mother-infant interactions." *Physiology & Behavior* 71: 43–49.

Mann, J., Smuts, B. 1999. "Behavioral Development in Wild Bottlenose Dolphin Newborns (*Tursiops sp.*)." *Behaviour* 136: 529–566.

McCormack, K., et al. 2015. "The Development of an Instrument to Measure Global Dimensions of Maternal Care in Rhesus Macaques (*Macaca Mulatta*)." *American Journal of Primatology* 77: 20–33.

Murray, L., et al. 2016. "The functional architecture of mother-infant communication, and the development of infant social expressiveness in the first two months." *Scientific Reports* 6: 39019.

Numan, M., Young, L.J. 2016. "Neural mechanisms of mother–infant bonding and pair bonding: Similarities, differences, and broader implications." *Hormones and Behavior* 77: 98–112.

Pawluski, J., et al. 2016. "Neuroplasticity in the maternal hippocampus: Relation to cognition and effects of repeated stress." *Hormones and Behavior* 77: 86–97.

Perani, C.V., Slattery, D.A. 2014. "Using animal models to study post-partum psychiatric disorders." *British Journal of Pharmacology* 171: 4539–4555.

Robinson, K.J., et al. 2015. "Maternal Oxytocin Is Linked to Close Mother-Infant Proximity in Grey Seals (*Halichoerus grypus*)." *PLoS ONE* 10(12): e0144577.

Saltzman, W., Maestripieri, D. 2011. "The neuroendocrinology of primate maternal behavior." *Progress in Neuro-Psychopharmacology & Biological Psychiatry* 35: 1192–1204.

Sanchez, M.M., et al. 2017. "Social buffering of stress responses in nonhuman primates: Maternal regulation of the development of emotional regulatory brain circuits." *Social Neuroscience* 10: 5, 512–526.

Slattery, D.A., Hillerer, K.M. 2016. "The maternal brain under stress: Consequences for adaptive peripartum plasticity and its potential functional implications." *Frontiers in Neuroendocrinology* 41: 114–128.

Stanton, M.A., et al. 2015. "Maternal Behavior and Physiological Stress Levels in Wild Chimpanzees (*Pan troglodytes schweinfurthii*)." *International Journal of Primatology* 36: 473–488.

Stolzenburg, D.S., Champagne, F.A. 2016. "Hormonal and non-hormonal bases of maternal behavior: The role of experience and epigenetic mechanisms." *Hormones and Behavior* 77: 204–210.

Weir, J.S., et al. 2008. "Dusky dolphin (*Lagenorhynchus obscurus*) subgroup distribution: are shallow waters a refuge for nursery groups?" *Canadian Journal of Zoology* 86: 1225–1234.

Winberg, J. 2005. "Mother and Newborn Baby: Mutual Regulation of Physiology and Behavior—A Selective Review." *Developmental Psychobiology* 47: 217–229.

Zhou, Q., et al. 2014. "The Mutual Influences between Depressed *Macaca fascicularis* Mothers and Their Infants." *PLoS ONE* 9(3): e89931.

10: MAMMAL MOMS AND MILK

Abbondanza, F.N., et al. 2013. "Variation in the Composition of Milk of Asian Elephants (*Elephas maximus*) Throughout Lactation." *Zoo Biology* 32: 291–298.

Acevedo, J., et al. 2016. "Offspring kidnapping with subsequent shared nursing in Antarctic fur seals." *Polar Biology* 39: 1225–1232.

Aubin, T., et al. 2015. "Mother Vocal Recognition in Antarctic Fur Seal *Arctocephalus gazella* Pups: A Two-Step Process." *PLoS ONE* 10(9): e0134513.

Barrett, L., et al. 2006. "The ecology of motherhood: the structuring of lactation costs by chacma baboons." *Journal of Animal Biology* 75: 875–886.

Bernstein, R.M., Hinde, K. 2016. "Bioactive Factors in Milk Across Lactation: Maternal Effects and Influence on Infant Growth in Rhesus Macaques (*Macaca mulatta*)." *American Journal of Primatology* 78: 838–850.

Brennan, A.J., et al. 2007. "The Tammar Wallaby and Fur Seal: Models to Examine Local Control of Lactation." *Journal of Dairy Science* 90: E66–E75.

Bridges, R.S. 2016. "Long-term alterations in neural and endocrine processes induced by motherhood in mammals." *Hormones and Behavior* 77: 193–203.

Charrier, I., et al. 2001. "Mother's voice recognition by seal pups." *Nature* 412: 873–874.

Charrier, I., et al. 2003. "Vocal signature recognition of mothers by fur seal pups." *Animal Behaviour* 65: 543–550.

Cheng, Y., Belov, K. 2017. "Antimicrobial Protection of Marsupial Pouch Young." *Front Microbial* 8: 354.

Deacon, F., et al. 2015. "Concurrent pregnancy and lactation in wild giraffes (*Giraffa camelopardalis*)." *African Zoology* 50: 4, 331–334.

Dettmer, A.M., et al. 2015. "Associations between Parity, Hair Hormone Profiles during Pregnancy and Lactation, and Infant Development in Rhesus Monkeys (*Macaca mulatta*)." *PLoS ONE* 10(7): e0131692.

East, M.L., et al. 2015. "Does lactation lead to resource allocation trade-offs in the spotted hyaena?" *Behavioral Ecology and Sociobiology* 69: 805–814.

Enjapoor, A.K., et al. 2017. "Hormonal regulation of platypus Beta-lactoglobulin and monotreme lactation protein genes." *General and Comparative Endocrinology* 242: 38–48.

Enjapoori, A.K., et al. 2014. "Monotreme Lactation Protein Is Highly Expressed in Monotreme Milk and Provides Antimicrobial Protection." *Genome Biology and Evolution* 6(10): 2754–2773.

Griffiths, K., et al. 2017. "Prolonged transition time between colostrum and mature milk in a bear, the giant panda, *Ailuropoda melanoleuca*." *Royal Society of Open Science* 2: 150395.

Hewavisenti, R.V., et al. 2016. "The identification of immune genes in the milk transcriptome of the Tasmanian devil (*Sarcophilus harrisii*)." *PeerJ* 4: e1569.

Hinde, K. 2009. "Richer Milk for Sons but More Milk for Daughters: Sex-Biased Investment during Lactation Varies with Maternal Life History in Rhesus Macaques." *American Journal of Human Biology* 21: 512–519.

REFERENCES

Hinde, K., Capitanio, J.P. 2010. "Lactational Programming? Mother's Milk Energy Predicts Infant Behavior and Temperament in Rhesus Macaques (Macaca mulatta)." *American Journal of Primatology* 72: 522–529.

Hinde, K., et al. 2013. "Brief Communication: Daughter Dearest: Sex-Biased Calcium in Mother's Milk Among Rhesus Macaques." *American Journal of Physical Anthropology* 151: 144–150.

Hinde, K., et al. 2015. "Cortisol in mother's milk across lactation reflects maternal life history and predicts infant temperament." *Behavioral Ecology* 26(1): 269–281.

Hinde, K., Milligan, L.A. 2011. "Primate Milk: Proximate Mechanisms and Ultimate Perspectives." *Evolutionary Anthropology* 20: 9–23.

Hofer, H., et al. 2016. "Trade-offs in lactation and milk intake by competing siblings in a fluctuating environment." *Behavioral Ecology* 27(5): 1567–1578.

Hood, W.R., et al. 2014. "Milk composition and lactation strategy of a eusocial mammal, the naked mole-rat." *Journal of Zoology* 293: 108–118.

Kastelein, R.A., et al. 2015. "Behavior and Body Mass Changes of a Mother and Calf Pacific Walrus (*Odobenus rosmarus divergens*) During the Suckling Period." *Zoo Biology* 34: 9–19.

Kurrupath, S., et al. 2012. "Monotremes and marsupials: Comparative models to better understand the function of milk." *Journal of Bioscience* 37(4): 581–588.

Lang, S.L.C., et al. 2005. "Individual variation in milk composition over lactation in harbour seals (*Phoca vitulina*) and the potential consequences of intermittent attendance." *Canadian Journal of Zoology* 83: 12.

Luque, S.P., et al. 2007. "Foraging behaviour of sympatric Antarctic and subantarctic fur seals: does their contrasting duration of lactation make a difference?" *Marine Biology* 152: 213–224.

Markham, A.C., et al. 2014. "Rank effects on social stress in lactating chimpanzees." *Animal Behaviour* 87: 195–202.

Oftedal, O.T. 2012. "The evolution of milk secretion and its ancient origins." *Animal* 6:3.

Osthoff, G., et al. 2017. "Milk composition of free-ranging red hartebeest, giraffe, Southern reedbuck and warthog and a phylogenetic comparison of the milk of African Artiodactyla." *Comparative Biochemistry and Physiology* Part A 204: 93–103.

Power, M.L., et al. 2016. "Patterns of milk macronutrients and bioactive molecules across lactation in a western lowland gorilla (*Gorilla gorilla*) and a Sumatran orangutan (*Pongo abelii*)." *American Journal of Primatology* 79: e22609.

Sauvé, C.C., et al. 2015. "Mother-pup vocal recognition in harbour seals: influence of maternal behaviour, pup voice and habitat sound properties." *Animal Behaviour* 105: 109–120.

Skibiel, A.L., et al. 2013. "The evolution of the nutrient composition of mammalian milks." *Journal of Animal Ecology* 82: 1254–1264.

Van Noordwijk, M.A., et al. 2013. "The Evolution of the Patterning of Human Lactation: a Comparative Perspective." *Evolutionary Anthropology* 22: 202–212.

Zhang, T., et al. 2016. "Changeover from signalling to energy-provisioning lipids during transition from colostrum to mature milk in the giant panda (*Ailuropoda melanoleuca*)." *Scientific Reports* 6: 3614.

Zhang, T., et al. 2015. "Changes in the Milk Metabolome of the Giant Panda (*Ailuropoda melanoleuca*) with Time after Birth—Three Phases in Early Lactation and Progressive Individual Differences." *PLoS ONE* 10(12): e0143417.

11: OPPORTUNISTIC ORPHANS AND MAGNANIMOUS MOMS

Baldovino, M.C., Di Bitetti, M.S. 2007. "Allonursing in Tufted Capuchin Monkeys (*Cebus nigritus*): Milk or Pacifier?" *Folia Primatologia* 79: 79–92.

Brandlová, K., et al. 2013. "Camel Calves as Opportunistic Milk Thefts? The First Description of Allosuckling in Domestic Bactrian Camel (*Camelus bactrianus*)." *PLoS ONE* 8(1): e53052.

De Bruyn, P.J.N., et al. 2009. "Prevalence of allosuckling behaviour in Subantarctic fur seal pups." *Mammalian Biology* 75: 555–560.

Engelhardt, S.C., et al. 2014. "Allosuckling in reindeer (*Rangifer tarandus*): Milk-theft, mismothering or kin selection?" *Behavioural Processes* 107: 133–141.

Engelhardt, S.C., et al. 2014. "Evidence of Reciprocal Allonursing in Reindeer, *Rangifer tarandus*." *Ethology* 121: 245–259.

Engelhardt, S.C., et al. 2016. "Allonursing in reindeer, *Rangifer tarandus*: a test of the kin-selection hypothesis." *Journal of Mammalogy* 97(3): 689–700.

Engelhardt, S.C., et al. 2016. "Allosuckling in reindeer (*Rangifer tarandus*): A test of the improved nutrition and compensation hypotheses." *Mammalian Biology* 81: 146–152.

Franco-Trecu, V., et al. 2010. "Allo-suckling in the South American fur seal (*Arctocephalus australis*) in Isla de Lobos, Uruguay: cost or benefit of living in a group?" *Ethology Ecology & Evolution* 22: 2, 143–150.

Gloneková, M., et al. 2016. "Stealing milk by young and reciprocal mothers: high incidence of allonursing in giraffes, *Giraffa camelopardalis*." *Animal Behaviour* 113: 113–123.

Jones, J.D., Treanor, J.J. 2008. "Allonursing and Cooperative Birthing Behavior in Yellowstone Bison, *Bison bison*." *Canadian Field-Naturalist* 122(2): 171–172.

MacLeod, K.J., Lukas, D. 2014. "Revisiting non-offspring nursing: allonursing evolves when the costs are low." *Biology Letters* 10: 20140378.

Maniscalco, J.M., et al. 2007. "Alloparenting in Steller sea lions (*Eumetopias jubatus*): correlations with misdirected care and other observations." *Journal of Ethology* 25: 125–131.

Olléová, M., et al. 2012. "Effect of social system on allosuckling and adoption in zebras." *Journal of Zoology* 288: 127–134.

Pitcher, B.J., Ahonen, H. 2011. "Allosuckling behavior in the Australian sea lion (*Neophoca cinerea*): An updated understanding." *Marine Mammal Science* 27(4): 881–888.

Pluháček, J., Bartošová, J. 2011. "A case of suckling and allosuckling behaviour in captive common hippopotamus." *Mammalian Biology* 76: 380–383.

Pluháček, J., et al. 2011. "A case of adoption and allonursing in captive plains zebra (*Equus burchellii*)." *Behavioural Processes* 86: 174–177.

Roulin, A. 2002. "Why do lactating females nurse alien offspring? A review of hypotheses and empirical evidence." *Animal Behaviour* 63: 201–208.

Tanaka, I. 2004. "Non-offspring nursing by a nulliparous pregnant female just before first parturition in free-ranging Japanese macaques." *Primates* 45: 205–206.

Tučková, V., et al. 2016. "Alloparental behaviour in Sinai spiny mice *Acomys dimidiatus*: a case of misdirected parental care?" *Behavioral Ecology and Sociobiology* 70: 437–447.

Weidt, A., et al. 2014. "Communal nursing in wild house mice is not a by-product of group living: Females choose." *Naturwissenschaften* 101: 73–76.

12: OTHER MOTHERS: ADOPTION AND ALLOMOTHERING

Atkinson, S.N., et al. 1996. "A Case of Offspring Adoption in Free-Ranging Polar Bears (*Ursus maritimus*)." *Arctic* 49: 1, 94–96.

Bădescu, I., et al. 2016. "Alloparenting is associated with reduced maternal lactation effort and faster weaning in wild chimpanzees." *Royal Society of Open Science* 3: 160577.

Dunham, N.T., Opere, P.O. 2016. "A unique case of extra-group infant adoption in free-ranging Angola black and white colobus monkeys (*Colobus angolensis palliatus*)." *Primates* 57: 187–194.

East, M.L., et al. 2009. "Maternal effects on offspring social status in spotted hyenas." *Behavioral Ecology* 20: 478–483.

Flatz, R., Gerber, L.R. 2010. "First Evidence for Adoption in California Sea Lions." *PLoS ONE* 5(11): e13873.

Hewlett, B.S., Winn, S. 2014. "Allomaternal Nursing in Humans." *Current Anthropology* 55: 2, 200–229.

Izar, P., et al. 2006. "Cross-Genus Adoption of a Marmoset (*Callithrix jacchus*) by Wild Capuchin Monkeys (Cebus libidinosus): Case Report." *American Journal of Primatology* 68: 692–700.

King, W.J., et al. 2015. "Adoption in Eastern Grey Kangaroos: A Consequence of Misdirected Care?" *PLoS ONE* 10(5): e025182.

Kishimoto, T., et al. 2014. "Alloparenting for chimpanzee twins." *Scientific Reports* 4: 6306.

Lecomte, N., et al. 2006. "Alloparental feeding in the king penguin." *Animal Behaviour* 71: 457–462.

Malenfant, R.M., et al. 2016. "Evidence of adoption, monozygotic twinning, and low inbreeding rates in a large genetic pedigree of polar bears." *Polar Biology* 39: 1455–1465.

Martins, W.P., et al. 2007. "A case of infant swapping by wild northern muriquis (*Brachyteles hypoxanthus*)." *Primates* 48: 324–326.

McNutt, J.W. 1996. "Adoption in African wild dogs, *Lycaon pictus*." *Journal of Zoology* 240: 163–173.

Olléová, M., et al. 2012. "Effect of social system on allosuckling and adoption in zebras." *Journal of Zoology* 288: 127–134.

Pelé, M., Petit, O. 2015. "Equal care for own versus adopted infant in tufted capuchins (*Sapajus* spp.)." *Primates* 56: 201–206.

Preston, S.D. 2013. "The Origins of Altruism in Offspring Care." *Psychological Bulletin* 139: 6, 1305–1341.

Riedman, M.L. 1982. "The Evolution of Alloparental Care and Adoption in Mammals and Birds." *The Quarterly Review of Biology* 57: 4, 405–435.

Riedman, M.L., Le Boeuf, B.J. 1982. "Mother-Pup Separation and Adoption in Northern Elephant Seals." *Behavioral Ecology and Sociobiology* 11: 203–215.

Roldán, M, Soler, M. 2011. "Parental-care parasitism: how do unrelated offspring attain acceptance by foster parents?" *Behavioral Ecology* 22: 679–691.

Sakai, M., et al. 2016. "A wild Indo-Pacific bottlenose dolphin adopts a socially and genetically distant neonate." *Scientific Reports* 6: 23902.

Stanford, C.B. 1992. "Costs and Benefits of Allomothering in Wild Capped Langurs (*Presbytis pileata*)." *Behavioural Ecology and Sociobiology* 30: 1, 29–34.

Stiver, K.A., Alonzo, S.H. 2010. "Alloparental care increases mating success." *Behavioral Ecology* 22: 206–211.

Thierry, B., Anderson, J.R. 1986. "Adoption in Anthropoid Primates." *International Journal of Primatology* 7:2, 191–216.

13: THE MEANING OF WEANING

Borries, C., et al. 2014. "The Meaning of Weaning in Wild Phayre's Leaf Monkeys: Last Nipple Contact, Survival, and Independence." *American Journal of Physical Anthropology* 154: 291–301.

Eckardt, W., et al. 2016. "Weaned age variation in the Virunga mountain gorillas (*Gorilla beringei beringei*): influential factors." *Behavioral Ecology and Sociobiology* 70: 493–507.

Fragaszy, D.M., Adams-Curtis, L.E. 1997. "Developmental Changes in Manipulation in Tufted Capuchins (*Cebus apella*) From Birth Through 2 Years and Their Relation to Foraging and Weaning." *Journal of Comparative Psychology* 111(2): 201–211.

Madden, J.R., et al. 2009. "Why do meerkat pups stop begging?" *Animal Behaviour* 78: 85–89.

Maestripieri, D. 1995. "Maternal encouragement in nonhuman primates and the question of animal teaching." *Human Nature* 6(4): 361–378.

Mainwaring, M.C. 2016. "The transition from dependence to independence in birds." *Behavioral Ecology and Sociobiology* 70: 1416–1431.

Mandalaywala, T.M., et al. 2014. "Physiological and behavioural responses to weaning conflict in free-ranging primate infants." *Animal Behaviour* 97: 241–247.

Murray, C.M., et al. 2014. "Early social exposure in wild chimpanzees: Mothers with sons are more gregarious than mothers with daughters." *Proceedings of the National Academy of Sciences* 111(51): 18189–18194.

Naguib, M., et al. 2010. "Mother is not like mother: Concurrent pregnancy reduces lactating guinea pigs' responsiveness to pup calls." *Behavioural Processes* 83: 79–81.

Paul. M., et al. 2014. "Selfish mothers? An empirical test of parent-offspring conflict over extended parental care." *Behavioural Processes* 103: 17–22.

Paul, M., et al. 2015. "Selfish mothers indeed! Resource-dependent conflict over extended parental care in free-ranging dogs." *Royal Society of Open Science* 2: 150580.

Riou, S., et al. 2012. "Parent-offspring conflict during the transition to independence in a pelagic seabird." *Behavioral Ecology* 23: 5, 1102-1107.

Smith, T.M., et al. 2017. "Cyclical nursing patterns in wild orangutans." *Scientific Advances* 3: e1501517.

Thompson, M.E., et al. 2012. "The energetics of lactation and the return to fecundity in wild chimpanzees." *Behavioral Ecology* 23(6): 1234–1241.

Trillmich, F., et al. 2016. "Patient Parents: Do Offspring Decide on the Timing of Fledging in Zebra Finches?" *Ethology* 122: 411–418.

Van Noordwijk, M.A., et al. 2013. "Multi-year lactation and its consequences in Bornean orangutans (*Pongo pygmaeus wurmbii*)." *Behavioral Ecology and Sociobiology* 67(5): 805–814.

14: MENOPAUSING MOMS AND GRACIOUS GRANDMAS

Alberts, S.C., et al. 2013. "Reproductive aging patterns in primates reveal that humans are distinct." *Proceedings of the National Academy of Sciences of the United States of America* 110(33): 13440–13445.

Brent, L.J.N, et al. 2015. "Ecological Knowledge, Leadership, and the Evolution of Menopause in Killer Whales." *Current Biology* 25: 746–750.

Bronikowski, A.M., et al. 2011. "Aging in the Natural World: Comparative Data Reveal Similar Mortality Patterns Across Primates." *Science* 331: 1325–1328.

Croft, D.P., et al. 2015. "The evolution of prolonged life after reproduction." *Trends in Ecology & Evolution* 30(7): 407–417.

Croft, D.P., et al. 2017. "Reproductive Conflict and the Evolution of Menopause in Killer Whales." *Current Biology* 27: 298–304.

Descamps, S., et al. 2006. "Best squirrels trade a long life for an early reproduction." *Proceedings of theRoyal Society B* 273: 2369–2374.

Field, J.M., Bonsall, M.B. 2017. "Evolutionary stability and the rarity of grandmothering." *Ecology and Evolution* 7: 3574–3578.

Foster, E.A., et al. 2012. "Adaptive Prolonged Postreproductive Life Span in Killer Whales." *Science* 337: 1313.

Franks, D.W., et al. 2016. "The significance of postreproductive lifespans in killer whales: a comment on Robeck, et al." *Journal of Mammalogy* 97(3): 906–909.

Hamel, S., et al. 2011. "Tradeoff between offspring mass and subsequent reproduction in a highly iteroparous mammal." *Oikos* 120: 690–695.

Hamel, S., et al. 2012. "Maternal allocation in bison: co-occurrence of senescence, cost of reproduction, and individual quality." *Ecological Applications* 22(5): 1628–1639.

Hayward, A.D., et al. 2014. "Early reproductive investment, senescence and lifetime reproductive success in female Asian elephants." *Journal of Evolutionary Biology* 27: 772–783.

Hoffman, C.L., et al. 2010. "Terminal investment and senescence in rhesus macaques (*Macaca mulatta*) on Cayo Santiago." *Behavioral Ecology* 21: 972–978.

Kim, P.S., et al. 2012. "Increased longevity evolves from grandmothering." *Proceedings: Biological Sciences* 279(1749): 4880–4884.

King, S.L., et al. 2016. "Maternal signature whistle use aids mother-calf reunions in a bottlenose dolphin, *Tursiops truncatus*." *Behavioural Processes* 126: 64–70.

King, W.J., et al. 2017. "Long-term consequences of mother-offspring associations in eastern grey kangaroos." *Behavioral Ecology and Sociobiology* 71: 77.

Lacreuse, A., et al. 2008. "Menstrual Cycles Continue into Advanced Old Age in the Common Chimpanzee (*Pan troglodytes*)." *Biology of Reproduction* 79: 407–412.

Lahdenperä, M., et al. 2012. "Severe intergenerational reproductive conflict and the evolution of menopause." *Ecology Letters* 15: 1283–1290.

Lahdenperä, M., et al. 2014. "Reproductive cessation and post-reproductive lifespan in Asian elephants and pre-industrial humans." *Frontiers in Zoology* 11: 54.

Lahdenperä, M., et al. 2016. "Short-term and delayed effects of mother death on calf mortality in Asian elephants." *Behavioral Ecology* 27(1): 166–174.

Maestripieri, D., Hoffman, C.L. 2011. "Chronic stress, allostatic load, and aging in nonhuman primates." *Development and Psychopathology* 23: 1187–1195.

Mann, J., Smuts, B.B. 1998. "Natal attraction: allomaternal care and mother-infant separations in wild bottlenose dolphins." *Animal Behaviour* 55: 1097–1113.

Nakamura, M., et al. 2014. "Brief Communication: Orphaned Male Chimpanzees Die Young Even After Weaning." *American Journal of Physical Anthropology* 153: 139–143.

Packer, C., et al. 1998. "Reproductive cessation in female mammals." *Nature* 392: 807–811.

Paterson, J.T., et al. 2016. "Patterns of age-related change in reproductive effort differ in the pre-natal and post-natal periods in a long-lived mammal." *Journal of Animal Ecology* 85: 1540–1551.

Robbins, A.M., et al. 2016. "Mothers may shape the variations in social organization among gorillas." *Royal Society of Open Science* 3: 160533.

Robinson, M.R., et al. 2012. "Senescence and age-specific trade-offs between reproduction and survival in female Asian elephants." *Ecology Letters* 15: 260–266.

Sefc, K.M., et al. 2012. "Brood mixing and reduced polyandry in a maternally mouthbrooding cichlid with elevated among-breeder relatedness." *Molecular Ecology* 21: 2805–2815.

Sharp, S.P., Clutton-Brock, T.H. 2010. "Reproductive senescence in a cooperatively breeding mammal." *Journal of Animal Ecology* 79: 176–183.

Tettamanti, F., et al. 2015. "Senescence in breeding success of female Alpine chamois (*Rupicapra rupicapra*): the role of female quality and age." *Oecologia* 178: 187–195.

Videan, E.N., et al. 2006. "The Effects of Aging on Hormone and Reproductive Cycles in Female Chimpanzees (*Pan troglodytes*)." *Comparative Medicine* 56(4): 291–299.

Walker, M.L., Herndon, J.G. 2008. "Menopause in Nonhuman Primates?" *Biology of Reproduction* 79: 398–406.

Wright, B.M., et al. 2016. "Kin-directed food sharing promotes lifetime natal philopatry of both sexes in a population of fish-eating killer whales, *Orcinus orca*." *Animal Behaviour* 115: 81–95.

15: MOTHERS IN MOURNING

Anderson, J.R. 2016. "Comparative thanatology." *Current Biology* 26: R543–R576.

Anderson, J.R., et al. 2010. "Pan thanatology." *Current Biology* 20(8): R349–R350.

Appleby, R., et al. 2013. "Observations of a free-ranging adult female dingo (*Canis dingo*) and littermates' responses to the death of a pup." *Behavioural Processes* 96: 42–46.

Bercovitch, F.B. 2012. "Giraffe cow reaction to the death of her newborn calf." *African Journal of Ecology* 51: 376–379.

Biro, D., et al. 2010. "Chimpanzee mothers at Bossou, Guinea carry the mummified remains of their dead infants." *Current Biology* 20(8): R351–R352.

Campbell, L.A.D., et al. 2016. "Behavioral responses to injury and death in wild Barbary macaques (*Macaca sylvanus*)." *Primates* 57: 309–315.

Caretta, J.V., et al. 1995. "A Risso's Dolphin (*Grampus Griseus*) Carrying a Dead Calf." *Marine Mammal Science* 11(4): 593–594.

Cronin, K.A., et al. 2011. "Behavioral Response of a Chimpanzee Mother Toward her Dead Infant." *American Journal of Primatology* 73: 415–421.

Douglas-Hamilton, I. 2006. "Behavioural reactions of elephants towards a dying and deceased matriarch." *Applied Animal Behaviour Science* 100: 87–102.

Fashing, P.J., et al. 2011. "Death Among Geladas (*Theropithecus gelada*): A Broader Perspective on Mummified Infants and Primate Thanatology." *American Journal of Primatology* 73: 405–409.

Hubbs, C.L., et al. 1953. "Dolphin Protecting Dead Young." *Journal of Mammalogy* 34(4): 498–518.

Li, T., et al. 2012. "Maternal responses to dead infants in Yunnan snub-nosed monkey (*Rhinopithecus bieti*) in the Baimaxueshan Nature Reserve, Yunnan, China." *Primates* 53: 127–132.

Majolo, B., McFarland, R. 2009. "Brief Communication: Self-Suckling in Barbary Macaque (*Macaca sylvanus*) Mothers Before and After the Death of Their Infant." *American Journal of Physical Anthropology* 140: 381–383.

Noren, S.R., Edwards, E.F. 2007. "Physiological and behavioral development in delphinid calves: Implications for calf separation and mortality due to tuna purse-seine sets." *Marine Mammal Science* 23(1): 15–29.

Payne, K. 2003. "Sources of Social Complexity in the Three Elephant Species." *Animal Social Complexity*.

Quintana-Rizzo, E., Wells, R.S. 2016. "Behavior of an Adult Female Bottlenose Dolphin (*Tursiops truncatus*) Toward an Unrelated Dead Calf." *Aquatic Mammals* 42(2): 198–202.

Reggente, M.A.L., et al. 2016. "Nurturant behavior toward dead conspecifics in free-ranging mammals: new records for odontocetes and a general review." *Journal of Mammalogy* 97(5): 1428–1434.

Sapolsky, R.M. 2016. "Psychiatric distress in animals versus animal models of psychiatric distress." *Nature Neuroscience* 19(11): 1387–1389.

Smith, T.G., Sleno, G.A. 1985. "Do white whales, *Delphinapterus leucas*, carry surrogates in response to early loss of their young?" *Canadian Journal of Zoology* 64: 1581–1582.

Strauss, M.K.L., Muller, Z. 2012. "Giraffe mothers in East Africa linger for days near the remains of their dead calves." *African Journal of Ecology* 51: 506–509.

Tokuyama, N., et al. 2017. "Cases of maternal cannibalism in wild bonobos (*Pan paniscus*) from two different field sites, Wamba and Kokolopori, Democratic Republic of the Congo." *Primates* 58: 7–12.

Van Leeuwen, E.J.C., et al. 2017. "Tool use for corpse cleaning in chimpanzees." *Scientific Reports* 7: 44091.

ADDITIONAL REFERENCES

Abbott, D.H., et al. 2015. "Nonhuman Primates Contribute Unique Understanding to Anovulatory Infertility in Women." *ILAR Journal* 45(2): 116–131.

REFERENCES

Bailey, A., et al. 2015. "Does pregnancy coloration reduce female conspecific aggression in the presence of maternal kin?" *Animal Behaviour* 108: 199–206.

Balshine, S., Sloman, K.A. 2011. "Parental Care in Fishes." In: Farrell A.P., (ed.), *Encyclopedia of Fish Physiology: From Genome to Environment, Volume 1*, pp. 670–677. San Diego: Academic Press.

Bardi, M., et al. 2001. "Hormonal Correlates of Maternal Style in Captive Macaques (Macaca fuscata and M. mulatta)." *International Journal of Primatology* 22(4): 647–662.

Bardi, M., et al. 2003. "Mother–infant relationships and maternal estrogen metabolites changes in macaques *(Macaca fuscata, M. mulatta)*." *Primates* 44: 91–98.

Bardi, M., et al. 2001. "Social Behavior and Hormonal Correlates during the Perinatal Period in Japanese Macaques." *Hormones and Behavior* 39: 239–246.

Bartoš, L., et al. 2011. "Promiscuous behaviour disrupts pregnancy block in domestic horse mares." *Behavioral Ecology and Sociobiology* 65: 1567–1572.

Beehner, J.C., et al. 2006. "The ecology of conception and pregnancy failure in wild baboons." *Behavioral Ecology* 17: 741–750.

Beehner, J.C., Lu, A. 2013. "Reproductive Suppression in Female Primates: A Review." *Evolutionary Anthropology* 22: 226–238.

Best, P.B., et al. 2015. "Possible non-offspring nursing in the southern right whale, Eubalaena australis." *Journal of Mammalogy* 96(2): 405–416.

Blomquist, G.E. 2013. "Maternal Effects on Offspring Mortality in Rhesus Macaques *(Macaca mulatta)*." *American Journal of Primatology* 75: 238–251.

Bothma, J. du P. 2004. "Motherhood Increases Hunting Success in Southern Kalahari Leopards." *Journal of Mammalogy* 85(4): 756–760.

Burkart, J.M. 2015. "Opposite effects of male and female helpers on social tolerance and proactive prosociality in callitrichid family groups." *Scientific Reports* 5: 9622.

Catalano, R.A., et al. 2016. "Twinning in Norway Following the Oslo Massacre: Evidence of a 'Bruce Effect' in Humans." *Twin Research and Human Genetics* 19(5): 485–491.

Charrier, I., et al. 2001. "Mother's voice recognition by seal pups." *Nature* 412: 873–874.

Clarke, M.R., et al. 1998. "Infant-Nonmother Interactions of Free-Ranging Mantled Howlers (Alouatta palliata) in Costa Rica." *International Journal of Primatology* 19(3): 451–472.

Curno, O., et al. 2009. "Mothers produce less aggressive sons with altered immunity when there is a threat of disease during pregnancy." *Proceedings of the Royal Society B* 276: 1047–1054.

De Lathouwers, M., Van Elsacker, L. 2005. "Reproductive Parameters of Female Pan paniscus and P. troglodytes: Quality versus Quantity." *International Journal of Primatology* 26(1): 55–71.

Dettmer, A.M., et al. 2016. "Associations between early life experience, chronic HPA axis activity, and adult social rank in rhesus monkeys." *Social Neuroscience* 12(1): 92–101.

Dettmer, A.M., et al. 2016. "First-Time Rhesus Monkey Mothers, and Mothers of Sons, Preferentially Engage in Face-to-Face Interactions with their Infants." *American Journal of Primatology* 78: 238–246.

REFERENCES

Dettmer, A.M., et al. 2016. "Neonatal face-to-face interactions promote later social behaviour in infant rhesus monkeys." *Nature Communications* 7: 11940.

Douglas, P.H., et al. 2016. "Mixed messages: wild female bonobos show high variability in the timing of ovulation in relation to sexual swelling patterns." *BMC Evolutionary Biology* 16:140.

Dunbar, R.I.M, Dunbar, E.P. 1977. "Dominance and reproductive success among female gelada baboons." *Nature* 266: 351–352.

Eckardt, W., et al. 2016. "Weaned age variation in the Virunga mountain gorillas (*Gorilla beringei beringei*): influential factors." *Behavioral Ecology and Sociobiology* 70: 493–507.

Erb, W.M., Porter, L.M. 2017. "Mother's little helpers: What we know (and don't know) about cooperative infant care in callitrichines." *Evolutionary Anthropology* 26: 25–37.

Fashing, P.J., Nguyen, N. 2011. "Behavior toward the dying, diseased, or disabled among animals and its relevance to paleopathology." *International Journal Paleopathology* 1: 128–129.

Grey, K.R., et al. 2013. "Human milk cortisol is associated with infant temperament." *Psychoneuroendocrinology* 38: 1178–1185.

Hackländer, K., Arnold, W. 1999. "Male-caused failure of female reproduction and its adaptive value in alpine marmots (*Marmota marmota*)." *Behavioral Ecology* 10: 592–597.

Johnson, R.L., Kapsalis, E. 1998. "Menopause in Free-Ranging Rhesus Macaques: Estimated Incidence, Relation to Body Condition, and Adaptive Significance." *International Journal of Primatology* 19(4): 751–765.

Kim, S., et al. 2014. "Maternal oxytocin response predicts mother-to-infant gaze." *Brain Research* 1580: 133–142.

Kinsley, C.H., et al. 2008. "Motherhood Induces and Maintains Behavioral and Neural Plasticity across the Lifespan in the Rat." *Archives of Sexual Behavior* 37: 43–56.

Kristal, M.B., et al. 2012. "Placentophagia in Humans and Nonhuman Mammals: Causes and Consequences." *Ecology of Food and Nutrition* 51(3): 177–197.

Kumar, A., et al. 2005. "Observations on parturition and allomothering in wild capped langur (*Trachypithecus pileatus*)." *Primates* 46: 215–217.

Li, T., et al. 2012. "Maternal responses to dead infants in Yunnan snub-nosed monkey (*Rhinopithecus bieti*) in the Baimaxueshan Nature Reserve, Yunnan, China." *Primates* 53: 127–132.

Lomanowska, A.M. 2017. "Parenting Begets Parenting: A Neurobiological Perspective on Early Adversity and the Transmission of Parenting Styles across Generations." *Neuroscience* 342: 120–139.

Maestripieri, D. 1999. "Changes in Social Behavior and Their Hormonal Correlates during Pregnancy in Pig-tailed Macaques." *International Journal of Primatology* 20(5): 707–718.

Maestripieri, D. 1994. "Social Structure, Infant Handling, and Mothering Styles in Group-Living Old World Monkeys." *International Journal of Primatology* 15(4): 531–553.

Martins, V., et al. 2015. "Parturition and potential infanticide in free-ranging Alouatta guariba clamitans." *Primates* 56: 119–125.

Medina, I., Langmore, N.E. 2016. "The evolution of acceptance and tolerance in hosts of avian brood parasites." *Biology Reviews* 91: 569–577.

Mendonça, R.S., et al. 2017. "Development and behavior of wild infant-juvenile East Bornean orangutans (Pongo pygmaeus morio) in Danum Valley." *Primates* 58: 211–224.

Merkling, T., et al. 2015. "Maternal effects as drivers of sibling competition in a parent–offspring conflict context? An experimental test." *Ecology and Evolution* 6(11): 3699–3710.

Moriceau, S., Sullivan, R.M. "Neurobiology of Infant Attachment." *Developmental Psychobiology* 47: 230–242.

Nakamichi, M., et al. 1996. "Maternal Responses to Dead and Dying Infants in Wild Troops of Ring-Tailed Lemurs at the Berenty Reserve, Madagascar." *International Journal of Primatology* 17(4): 505–523.

Otali, E., Gilchrist, J.S. 2006. "Why chimpanzee (Pan troglodytes schweinfurthii) mothers are less gregarious than nonmothers and males: the infant safety hypothesis." *Behavioral Ecology and Sociobiology* 59: 561–570.

Paquet, M., Smiseth, P.T. 2016. "Maternal effects as a mechanism for manipulating male care and resolving sexual conflict over care." *Behavioral Ecology* 27(3): 685–694.

Pelé, M., Petit, O. 2015. "Equal care for own versus adopted infant in tufted capuchins (*Sapajus* spp.)." *Primates* 56: 201–206.

Ratnayeke, A.P., Dittus, W.P.J. 1989. "Observation of a Birth Among Wild Toque Macaques (Macaca sinica)." *International Journal of Primatology* 10(3): 235–242.

Reitsema, L.J., et al. 2016. "Inter-Individual Variation in Weaning Among Rhesus Macaques (*Macaca mulatta*): Serum Stable Isotope Indicators of Suckling Duration and Lactation." *American Journal of Primatology* 78: 1113–1134.

Ren, B., et al. 2012. "Evidence of Allomaternal Nursing across One-Male Units in the Yunnan Snub-Nosed Monkey (*Rhinopithecus Bieti*)." *PLoS ONE* 7(1): e30041.

Robinson, K.J., et al. 2015. "Maternal Oxytocin Is Linked to Close Mother-Infant Proximity in Grey Seals (*Halichoerus grypus*)." *PLoS ONE* 10(12): e0144577.

Rothe, H., et al. 1993. "Long-Term Study of Infant-Carrying Behavior in Captive Common Marmosets (Callithrix jacchus): Effect of Nonreproductive Helpers on the Parents' Carrying Performance." *International Journal of Primatology* 14(1): 79–93.

Setchell, J.M., et al. 2002. "Reproductive Parameters and Maternal Investment in Mandrills (*Mandrillus sphinx*)." *International Journal of Primatology* 23(1): 51–68.

Saltzman, W., Maestripieri, D. 2011. "The neuroendocrinology of primate maternal behavior." *Progress in Neuro-Psychopharmacology & Biological Psychiatry* 35: 1192–1204.

Santos, C.V., et al. 1997. "Infant Carrying Behavior in Callitrichid Primates: Callithrix and Leontopithecus." *International Journal of Primatology* 18(6): 889–907.

Stanton, M.A. 2015. "Maternal Behavior and Physiological Stress Levels in Wild Chimpanzees (*Pan troglodytes schweinfurthii*)." *International Journal of Primatology* 36: 473–488.

Steinetz, B.G., et al. 1997. "Pattern and Source of Secretion of Relaxin in the Reproductive Cycle of the Spotted Hyena (Crocuta crocuta)." *Biology of Reproduction* 56: 1301–1306.

REFERENCES

Takahata, Y., et al. 2001. "Daytime Deliveries Observed for the Ring-tailed Lemurs of the Berenty Reserve, Madagascar." *Primates* 42(3): 267–271.

Thompson, M.E. 2013. "Comparative Reproductive Energetics of Human and Non-human Primates." *Annual Review of Anthropology* 42: 287–304.

Ware, R.W., Catenazzi, A. 2016. "Pouch brooding marsupial frogs transfer nutrients to developing embryos." *Biology Letters* 12(10): 1–4.

ACKNOWLEDGMENTS

I am grateful every day for the generous support of many wonderful people. I thank my fellow allomoms who provide extra maternal care during the times when I'm busy with alternative mothering tasks: Karen, Vanessa, Cathy, Brenda, Chelsea, Chelsey, Katie, Debbie, Jen, Lisa, and many others. I'm grateful for the supportive and true friendship of Amber and Mary, and for the dedication and hard work of Andrea Meyer and Pawel Wojcikowski.

I thank all of my photo contributors, who graciously and generously offered up their wares for display in this book: Unding Jami, Yoram Shpirer, Kathy West, Adriana Lowe, Brock Fenton, Michelle LaRue, Joe Chowaniec, John King II, and Sarah Sovereign.

Lastly, I humbly acknowledge the support and friendship of my Pegasus editor Iris Blasi, who reached out to me about this book before I even realized that I should write it <3.

ABOUT THE AUTHOR

Carin Bondar holds a PhD in freshwater ecology from the University of British Columbia. She has written and hosted a variety of programs on TV and online (The Science Channel, Discovery Networks worldwide, National Geographic Wild, and Scientific American). She is the writer and host of the animated online series *Wild Sex* (based on her last book of the same name—with over 30 million views to date) and *Wild Moms* (Spring 2018—both at Seeker) .

In addition to her writing and hosting work, Bondar works with Taxon Expeditions (Leiden, Netherlands) to bring citizen scientists to remote ecosystems to discover new species. When she is not working, Bondar loves to make glass art, bake, go on long hikes, and spend time with her 4 children—Shaeden, Loanna, Faro, and Juna—her 3 dogs, and her cat.

INDEX

OCT 3 1 2019

CPSIA information can be obtained
at www.ICGtesting.com
Printed in the USA
LVHW081550201019
634740LV00005B/73/P

9 781643 132327